Disaster Planning for the Clinical Practice

Neil Baum, MD
Clinical Associate Professor of Urology
Tulane School of Medicine
Louisiana State University
New Orleans, Louisiana

John W. McDaniel, MHA
President and Chief Executive Officer
Peak Performance Physicians, LLC
New Orleans, Louisiana

JONES AND BARTLETT PUBLISHERS
Sudbury, Massachusetts
BOSTON   TORONTO   LONDON   SINGAPORE

World Headquarters
Jones and Bartlett
   Publishers
40 Tall Pine Drive
Sudbury, MA 01776
978-443-5000
info@jbpub.com
www.jbpub.com

Jones and Bartlett
   Publishers Canada
6339 Ormindale Way
Mississauga, Ontario
L5V 1J2
Canada

Jones and Bartlett
   Publishers International
Barb House, Barb Mews
London W6 7PA
United Kingdom

Jones and Bartlett's books and products are available through most bookstores and online booksellers. To contact Jones and Bartlett Publishers directly, call 800-832-0034, fax 978-443-8000, or visit our website www.jbpub.com.

Substantial discounts on bulk quantities of Jones and Bartlett's publications are available to corporations, professional associations, and other qualified organizations. For details and specific discount information, contact the special sales department at Jones and Bartlett via the above contact information or send an email to specialsales@jbpub.com.

Copyright © 2009 by Jones and Bartlett Publishers, LLC
Cover Image © Carolina K. Smith, M.D./ShutterStock, Inc.

All rights reserved. No part of the material protected by this copyright may be reproduced or utilized in any form, electronic or mechanical, including photocopying, recording, or by any information storage and retrieval system, without written permission from the copyright owner.

This publication is designed to provide accurate and authoritative information in regard to the subject matter covered. It is sold with the understanding that the publisher is not engaged in rendering legal, accounting, or other professional service. If legal advice or other expert assistance is required, the service of a competent professional person should be sought.

Library of Congress Cataloging-in-Publication Data

Baum, Neil, 1943-
  Disaster planning for the clinical practice / by Neil Baum and John W. McDaniel.
     p. ; cm.
  Includes bibliographical references and index.
  ISBN-13: 978-0-7637-5073-2 (pbk.)
  ISBN-10: 0-7637-5073-5 (pbk.)
  1. Medicine–Practice. 2. Emergency management. I. McDaniel, John W. II. Title.
  [DNLM: 1. Practice Management, Medical–organization & administration.
  2. Disaster Planning–organization & administration. 3. Disasters. W 80 B347d 2009]
  R728.B35 2009
  610.68–dc22
                                                                      2008008857

6048

Production Credits
Publisher: David Cella
Editorial Assistant: Maro Asadoorian
Production Director: Amy Rose
Production Editor: Renée Sekerak
Associate Marketing Manager: Lisa Gordon
Manufacturing and Inventory Control Supervisor: Amy Bacus
Cover Design: Anne Spencer
Composition: Auburn Associates, Inc.
Printing and Binding: Malloy Incorporated
Cover Printing: Malloy Incorporated

Printed in the United States of America
12 11 10 09 08   10 9 8 7 6 5 4 3 2 1

This book is dedicated to the doctors, nurses, and other health professionals who were responsible for maintaining the health of the citizens of the Gulf Coast area after Hurricane Katrina. Your heroic efforts helped to rebuild our communities and return the area to its historic splendor.

Contents

Acknowledgments................................. ix
Introduction..................................... xi

Chapter 1 **Disasters Come in Many Shapes and Sizes** 1
Categories of Disasters 3

Chapter 2 **Technological Disasters** 9
Protecting Your Computer from Unwanted Eyes
    and Cyberthieves 12
Cyberthieves 16
Responses to Spyware 17
Protecting Against Spyware and Adware 18
Passwords to Protect Your Computers.............. 19
Firewalls 21
Encryption...................................... 22
Surge Protection................................ 24
Levels of Surge Protection 24
Data Removal and Erasing Hard Drives 26
Impact of Fire on Data Loss 27
Choosing a Fire Protection Solution 28
Fire Detection System Types 29
Fire-Suppression System Types 31
Pull Stations 35
Signaling Devices............................... 36
Control Systems................................ 36
Protecting Mission-Critical Facilities.............. 37

Industry Best Practices 37
Common Mistakes 40
Bottom Line 40

Chapter 3 **Preparing a Disaster Plan** 43
Preparation of a Disaster Plan..................... 44
Human Resources 50
Key Contacts...................................... 53
Practice Operations............................... 55
Computer Equipment and Software 59
Voice/Data Communications 62
Miscellaneous Resources 66
Disaster Response Checklist 67
Do Your Employees Know About Your
   Emergency Plans? 69
Vital Records..................................... 69
Are Your Employees Prepared at Home? 70
Your Building 71
Practice Makes Perfect........................... 72
Getting Started................................... 73
Bottom Line 75

Chapter 4 **Implementing the Practice Resumption Plan** 77
Actions to Take After a Disaster 77
Implementation Plan 81
Prevention 81
Plan Development Checklists 82
Patient List 83
Practice Recovery Steps........................... 84
Guidelines for Travel to the Practice Recovery
   Site 86
Offsite Stored Materials 87
Critical Resources to Be Retrieved 87
Training/Exercise Schedule 88
Resumption Activities for Information System
   Capabilities................................... 90
Bottom Line 91

Chapter 5	**Protecting and Recovering Practice Assets** 93	
	Bottom Line 103	
Chapter 6	**Creating a Backup Plan** 105	
	Patient Demographic and Insurance Information 109	
	Practice Financial Information 110	
	Clinical Information 110	
	How Can Your Practice Back Up Important Data? 110	
	What Is the Cost of Backing Up Your Data? 112	
	How to Select the Correct Data Protection Solution. 112	
	Types of Data Backup 112	
	Which Individuals Inside and Outside of Your Practice Can Provide Assistance for a Data Backup System? 113	
	It Isn't Enough to Have a Plan—You Must Test That Plan. 119	
	Other Resources for Backing Up Your Data. 121	
	Bottom Line 121	
Chapter 7	**Before and After a Disaster: A Hospital Perspective** 125	
	Before a Disaster 128	
	What Can Your Hospital Do for You and Your Practice? 130	
	Bottom Line 131	
Chapter 8	**Insurance for Ameliorating the Pain of a Disaster** 133	
	Selecting Disaster Insurance 134	
	Property Insurance 136	
	Business Interruption Insurance 137	
	Extra Expense Insurance. 141	
	Bottom Line 141	

Chapter 9	**Finding an Alternative Site for Your Practice** ...143	
	Calculate the Cost of Down Time 145	
	Alternate Site Development Plan 149	
	Bottom Line 150	
Chapter 10	**Disaster Planning for the Employees** 153	
	Build a Disaster Kit 158	
	Pets 161	
	Food 161	
	Water 162	
	Utilities 162	
	Evacuation 164	
	Fire 165	
	Disaster Shelters 165	
	Bottom Line 166	
Chapter 11	**Conclusion** 167	

Appendix 1–19 171
Index 221

Acknowledgments

How does one thank others for helping with a book about disasters? Could anyone possibly imagine that something positive comes from a disaster such as Hurricane Katrina, which devastated the Gulf Coast Region in August 2005? Out of that disaster, however, we made a decision to increase others' awareness of disasters, both natural and man-made, as our experience could have been mitigated or significantly reduced if a disaster plan had been in place. We wrote this book hoping that others, without going through a disaster, can learn from our experience and not duplicate our mistakes.

We thank our publisher, Jones and Bartlett, and their staff, David Cella, Lisa Gordon, and Maro Asadoorian, who have been cooperative and helpful in making this book a reality.

To Les Hirsch, the CEO at Touro Infirmary, Frank Folino, the Vice President of Diagnostic and Ancillary Services, and Hal Leftwich, DBA, Chief Executive Officer, Hancock Medical Center, Bay St. Louis, MS, who truly weathered the storm and made their hospitals function in a short period of time after the storm passed.

Marjorie Satinsky, the president of Satinsky Consulting, was very helpful in contributing to the chapter on insurance.

Mr. Stan Levenson, founder of Levenson and Hill Public Relations firm in Dallas, Texas, edited this manuscript. He has been a good friend, a mentor, and valuable resource for all of my books.

Benjamin A. Swig is obtaining a masters in public health at Tulane University School of Public Health and Tropical Medicine in

Environment Health concentrating in disaster management. He was a great resource and provided invaluable assistance in editing and reviewing several chapters.

No book on disaster planning would be complete without thanking those brave first responders from the police, fire departments, and the U.S. Coast Guard, who flew countless missions and saved thousands of lives. These courageous people helped to take thousands out of harm's way after Hurricane Katrina.

Neil Baum and John McDaniel

Introduction

WHY PREPARE FOR A DISASTER?

Imagine that a disastrous storm is forecast and that you are advised to leave your community as soon as possible. Thus, you close your office, pack a few items, get your family, and then drive or fly out of town. Now you are out of harm's way. By attentively following the news reports, you learn that your community is under 8 to 10 feet of water and that the National Guard is called to restore law and order. You do not know what happened to your home, practice, or hospital. After looking at satellite photos on the Internet, you can see that the rooftops of your home and office are still attached to the foundations; nevertheless, flood water is everywhere. You do not know when you can safely return to your community to assess the damage. Your staff has scattered to safety, but you do not know how to reach them, as their cell phones are no longer functional. Patients are beginning to send you e-mails asking for medication refills, results of tests and studies, and copies of their medical records; they want to know where they should go for medical care.

That very scenario happened to me (NHB) on August 29, 2006, when Hurricane Katrina devastated New Orleans. Although no formal disaster plan was in place, I did have an electronic medical record program and was able to respond to and help care for my patients through phone conversations and the use of the Internet.

After living in Texas for nearly 11 weeks, I was able to return to New Orleans and rebuild my practice. Many of the problems and obstacles I faced could have been eliminated or minimized if I had had a disaster plan. After experiencing the anxiety and the costs of trying to rebuild my practice, to maintain my staff, and to provide continuity of care for my patients, we decided that it would be a service to our colleagues to share strategic guidelines about how to prepare and implement a disaster plan. Hopefully, all who read this book will take action and diligently prepare a disaster plan for your practices. For those who do develop and manage a disaster plan, you can be sure that you will not experience the apprehension and far-reaching impact that accompanies a disaster.

Any kind of "planning" related to "disasters" may seem the work of pessimists; however, in the aftermath of September 11, 2001, and August 29, 2005, when Hurricane Katrina devastated the Gulf Coast, certainly every medical practice must have the necessary plans to guide their activities if even the most unforeseen events occur. Under any circumstance, practices need to be able to ensure the safety of their employees and to continue to care for their patients.

This book is intended to provide guidelines and suggestions to put your disaster plan in place. Forms are available for download from the accompanying CD. These can then be distributed to various staff members to complete the preparation of your disaster plan.

Maintenance of your predisaster staff is critical to the success of a disaster plan. It is a daunting task to rebuild your practice even with your existing staff. Trying to rebuild with a new staff (even if only a few are new) can create an entirely new level of obstacles and challenges. Maintaining your staff requires timely and effective communication with your employees during a time of crisis. This book explains the techniques and methods that are necessary to keep your staff informed, and it clarifies their responsibilities in making the practice functional as quickly as possible.

In Chapter 1, we review the different types of disasters that can impact your practice. For example, the disaster preparation for family practice would be different from an ophthalmology practice, which contains expensive and delicate equipment. Also, geographic considerations are important. A practice in California, which is concerned about earthquakes, would require different preparations than a practice in Florida, which is at risk for hurricanes and accompanying wind and water damage.

Chapter 2 focuses on technological disasters, probably the most common disaster that affects medical practices, and how to prevent them. Technological failures, not natural disasters, are the most frequent and can cause as much or more havoc. Fortunately, most of these computer disasters can be prevented, and if they occur, the down time can be shortened considerably by proper preparation. This chapter explains the necessary steps in preventing computer disasters and how to get computers working again as quickly as possible.

In Chapter 3, we discuss the creation of a disaster plan using the team approach. When the time comes for everyone to launch into action, there will be a sense of camaraderie; this team spirit will help make the process of rebuilding much more effective and efficient. You will see that it is much easier to create your plan and assign responsibilities *before* the disaster rather than after.

In Chapter 4, we cover the implementation process of your disaster plan. After all, having the best disaster drill ideas will not help unless everyone knows the plan and how to execute it. This chapter explains the steps that need to take place before the disaster occurs and creates the roadmap for the recovery after the disaster.

Chapter 5 reviews how to protect and recover your vital assets. Having quick access to your data will determine the success of your practice. The protection and recovery of key medical practice assets will ensure that the practice is able to resume operations quickly and seamlessly after a disruptive event.

Chapter 6 provides suggestions for backing up your important data. Unfortunately, most medical practices and other non health care professionals do not back up their data consistently and reliably. Loss of this data can severely impact your practice; however, if you adhere to the advice in this chapter, rest assured that you will have timely access to your data. Today, most practices are dependent on technology, especially electronics, in order to be fully functional. Chapter 6 covers the process of backing up and retrieving electronic data and discusses various options for storing electronic data, including, but not limited to, backup tapes and Internet storage of electronic data. The steps to retrieving data from damaged backup tapes, discs, and hard drives are also covered in this chapter.

Chapter 7 discusses your hospital and hospital board's point of view. A disaster that affects your hospital will impact the care of your patients if they require hospitalization. Nearly every hospital has its own disaster plan. In this chapter, we discuss how you can dovetail onto the hospital's plan and make use of aspects of their plan for your practice.

Chapter 8 addresses the insurance coverage that is needed to protect your practice. It also identifies the types of business interruption insurance a practice needs in case of disaster. The pros and cons of buying extra-expense coverage are also discussed.

On occasion, you may have to leave the building where you practice because of extensive damage. Chapter 9 discusses the preparation of an alternative site to practice and how to get the practice functioning within 48 hours or less after a disaster occurrence.

Chapter 10 emphasizes that employees need to have their own personal disaster plan. Even with a thorough and effective disaster plan, however, you will not be able to implement your practice plan if your employees are not able to return to work; therefore, you must make every effort to motivate all of your employees to have a disaster plan that takes care of their home and their families. This chapter covers disaster planning for your staff, including office managers and physicians. It provides suggestions that your staff members will need to know in order to create their own disaster plan.

Thus, who should read this book? If you are in the health care profession and are concerned about the continuity of care for your patients when a disaster occurs, then you will want to read this book. Disasters are not confined to the Gulf Coast and eastern seaboard of the United States. Every practice is at a potential risk for either a natural or man-made disaster, such as a power outage and complete loss of your electronic data.

In conclusion, disasters are always costly. Frequently, people think of highly popularized, large-scale disasters seen on television as the costliest disasters; however, the common costs associated with everyday disasters such as fire, water damage, and loss of electronic data are far more frequent. It has long been established that electrical power outages, surges, and spikes cause more than $150 billion in damages to the U.S. economy every year. But what we fail to recognize is that 80 percent of power problems occur within an organization's own electrical power infrastructure . . . not because of external causes, for example, their public utility.[1]

Not only does a disaster result in loss of continuity of patient care, but it also impacts the loss of income and family support to your employees. Patients may have to seek advice from a neighboring practice if their regular doctor's practice is not open and is being restored. Patients will be upset and may even feel abandoned, as they will have to reschedule elective appointments, seek alternative advice midway through therapy (this could be very difficult for a patient undergoing chemo cancer therapy where the records of the original physician/oncologist cannot be found). Business disruption may also affect patient loyalty and a practice's reputation. Hopefully, after reading this book, you will mitigate these events and perhaps make your practice more disaster resistant.

None of our practices is impervious to a disaster or an emergency. An emergency can happen unexpectedly at any time. A disaster does not coincide with the busiest time of year for your practice or give you the luxury of occurring during a slow period; however, if you plan ahead and prepare, be assured that your patients will have access to your practice and their records and that you can jump start your practice in a timely fashion and either rebuild your existing practice or, if necessary, move to another facility and start all over. With proper insurance, you will be protected against an economic disaster.

BOTTOM LINE

Now is the time to prepare your practice for what could happen. Whether the calamity is widespread, affecting the entire community or region, or focused solely on your practice, you need to have a disaster plan in place. Hopefully, this book provides motivation and insight to initiate the steps necessary to create an effective plan of action. There is no way to completely bulletproof your practice from a disaster; however, you can lessen the impact, the losses, and the heartache that accompany a disaster. We hope that you will never need to use the plan but that you will have the peace of mind that you are prepared to weather any storm!

REFERENCE

1. Neumann, J. R. "Power Analytics and Business Continuity." Available at: http://www.edsa.com/infoCenter/pdfs/Power_Analytics_and_Business_Continuity_1_.pdf, accessed January 27, 2008.

Chapter 1

Disasters Come in Many Shapes and Sizes

There's no disaster that can't become a blessing, and no blessing that can't become a disaster.

—Richard Bach from *Jonathan Livingston Seagull*

Dr. Robert Kotler—a plastic surgeon in Beverly Hills, California, and author of *The Essential Cosmetic Surgery Companion: Don't Consult a Cosmetic Surgeon Without This Book!*—had hundreds of before and after patient photographs stored in his computer. These, part and parcel of his art style, were shown to his new patients to encourage them to proceed with cosmetic surgery; however, his computer was stolen, and because the photographs were not backed up, he lost priceless photographs, thus severely impacting his business.

Before beginning this chapter, the following important questions[1] should clarify why it is important to develop an effective disaster plan and should help you to determine your ability to survive some common disaster scenarios.

Key Questions

1. If all of your office computers were stolen, do you have their serial numbers, original costs, and values, as well as the ability to recreate all patient and practice management data?

2. If your office was completely destroyed by fire, how long would it take you to contact your patients so that you can recreate all computer data, to contact your insurance company, to process claims, to contact referring physicians, and to generally get your practice operational again? Who in your practice is responsible for performing each of these functions?

3. If you had a heart attack, are your charts organized so that another physician or colleague could enter the practice without your patients suffering from any lapse in care?

4. If you suddenly could not come into the office, is a person designated to take over your patients care? If you have a partner, how much does he or she know about the day-to-day operations of your practice?

5. If you are unable to come into the office for a few days or weeks, could anyone locate anything on your desk or in your patients' charts? Is the answer the same if your office manager was sick or away at the same time?

6. If your receptionist, nurse, billing clerk, or office manager suddenly quit, do you know his or her filing system so that you can find information in his or her desk or on his or her computer? Do you have copies or know where he or she keeps the keys for the filing cabinets, the supplies, and the locked medications? Do you know all of the employees' respective passwords (including voice mail, computer log-in, e-mail, the software practice management program, and any other software applications that are routinely used and necessary to run the practice)?

7. If one of your staff members disappeared with patients' charts, would you have sufficient records to determine what was taken and when?

8. If your partner was suddenly disabled, do you have the manpower and the skills to manage until he or she returns? Do you have a contingency plan to manage such a situation?

9. If you or a partner in your practice was disabled for an extended period of time, will you or he or she be able to draw a salary? If so, how much and for how long? If you are a solo practitioner, how will expenses of the practice be paid while you are unable to generate income?

10. If you were to die or become completely and permanently disabled, would your receptionist, partners, or office manager be able to assist the person who is to assume responsibility of your patients? What burdens would this place on your partners and their spouses?

If you are unable to answer all of these questions quickly or satisfactorily, then you need to do some disaster recovery planning. Regardless of the size of your practice, an easily implemented plan needs to be created that will assist you or anyone in your practice if an unexpected interruption occurs. This chapter describes the various kinds and likelihood of disasters that can impact your practice and the important of preplanning.

CATEGORIES OF DISASTERS

Three major recognized categories of widespread disasters exist: (1) natural disasters, (2) accidents, and (3) intentional acts of violence. Of course, earthquakes, floods, hurricanes, fires, tornados, volcanic eruptions, heat waves, and cold snaps are the most common natural disasters and receive the most media attention. The most serious accidents are nuclear accidents, chemical or biological contamination, agricultural or industrial accidents involving insecticides or pesticides, and transportation accidents involving large vehicles such as planes, trains, and buses. Intentional acts of violence can be physical, such as a bombing or a shooting, or can involve the use of "terrorism agents" such as biological or chemical agents that cause bacterial or viral infections. Terrorists could potentially use the following biological agents: anthrax, cholera, plague, tularemia, and Q fever. Viral agents of concern include smallpox, Venezuelan equine encephalitis, and viral hemorrhagic fevers. Biological toxins include botulinum and staphylococcal enterotoxin B. Chemical agents attack specific body functions or organ systems. They include nerve agents such as sarin, various insecticides and pesticides, blister agents such as mustard and lewisite, choking agents such as phosgene and chlorine, blood-altering compounds such as hydrogen cyanide and cyanogen chloride, and riot-control gases that cause vomiting and eye tearing. Certainly dealing with chemical and bacterial agents is beyond the scope of this book, but the principles of disaster preparation are similar to natural disasters and serious accidents.[2]

Many times we think of disasters as a fire, a hurricane, or a flood that impacts or shuts down both a practice and a community. The most common disasters, however, are those that impact a single practice. These include a broken water pipe, a sprinkler that inappropriately floods the office, a computer crash, a power surge that shuts down the computers, or a power outage that renders the electronics inoperable. Other examples include theft by a disgruntled employee or a cleanup crew, backup tape loss, bioterrorism, or a terrorist attack.

Figure 1-1 shows that man-made disasters are far more common than natural disasters. As a result, no practice is immune to the risk(s) of a disaster that can severely impact and render a practice inoperable or severely handicapped.

After the horrific events of September 11, 2001, many small commercial businesses in the area surrounding the World Trade Center Towers went out of business. Direct losses did not force their demise, as most of these small businesses had property insurance; however, they were not able to withstand the loss of income as a result of closing their doors for an extended period of time. In other words, few of those commercial businesses had business interruption insurance. This same scenario can also apply to medical practices.

At a 2005 conference on emergency planning sponsored by the Mission, Kansas-based SkillPath (http://www.skillpath.com/seminfo.html/st/CONDREC), an organization that offers business training for executives in the United States, Canada, Australia, New Zealand,

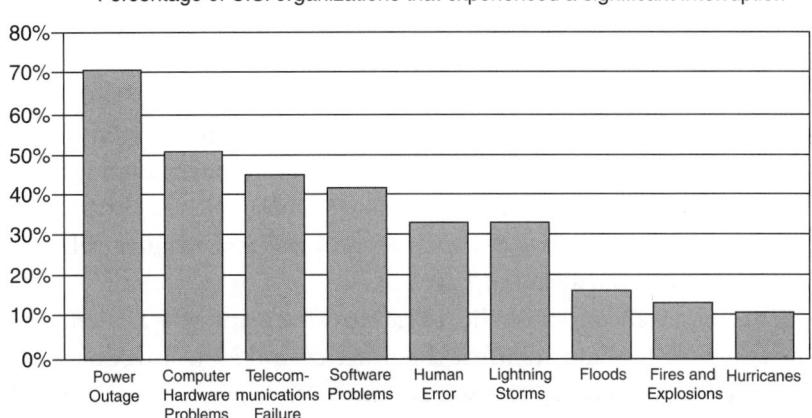

Figure 1-1 Disasters Happen Every Day.
Data from: Zinkewicz, P. "Business interruption insurance—death protection for a business." *Rough Notes* (Jul 2005): 1–3. Available at: http://findarticles.com/p/articles/mi_qa3615/is_200507/ai_n14775710, accessed January 29, 2008

and the United Kingdom, the following sobering statistics about U.S. companies were introduced:

- Seventy-two percent have had business operations significantly interrupted because of power outage.
- Fifty-two percent have had business operations significantly interrupted because of computer hardware problems.
- Forty-six percent have had business operations significantly interrupted because of telecommunications failure.
- Forty-three percent have had business operations significantly interrupted because of software problems.
- Thirty-four percent have had business operations interrupted because of human error.
- Thirty-four percent have had business operations interrupted because of lightning storms.
- Seventeen percent have had business operations significantly interrupted because of floods.
- Fourteen percent have had business operations significantly interrupted because of fires and/or explosions.

- Twelve percent have had business operations significantly interrupted because of hurricanes.

We tend to think that interruptions to business stem from catastrophic occurrences such as the World Trade Center tragedy or serious hurricanes; these statistics, however, clearly demonstrate that the highest percentages of business operation losses come from more mundane occurrences such as power outages and problems with computer hardware or software.

Unfortunately, most medical practices have an ostrich mentality about disasters that can affect the practice and even shut down the practice from a few hours to weeks and sometimes permanently. A survey by Media, Pennsylvania-based International Communications Research on behalf of Trusted Choice, the consumer marketing brand for agencies that are members of the Independent Insurance Agents and Brokers of America, showed that many U.S. small business owners, including small medical practices, have a major gap in insurance coverage, which pays for lost income if a practice is temporarily closed or if its income is reduced because of a covered loss. It can also pay for operating expenses (such as rent, salaries, and operating at a new location) that continue after business has come to a temporary halt.

The survey of 500 owners of small businesses—those with fewer than 100 employees and less than $1.5 million in revenue—shows that many have inadequate insurance coverage. Many medical practices are confused about what insurance protection is needed or required to return to a functioning practice in the face of a disaster.

A practice that has to close temporarily while its premises are being repaired may lose to competitors; may lose market share, employees, suppliers, and inventory; and may face other challenges. A quick resumption of operations is essential if a medical practice is to survive after a disastrous loss, and this can be accomplished in large part through business interruption coverage.[3]

Nobody likes to think about disasters, and most physicians have an attitude of "it won't happen to me"—they believe that disasters happen to the other guy.

The bottom line is this: Many medical practices will encounter a disaster that may be a natural disaster, an accident, or an intentional act of violence. Unfortunately, few practices are prepared for such a disaster. The next chapter discusses disaster planning preparation, what is important to do before the disaster occurs, and how to prepare for the disaster so that the impact is lessened.

REFERENCES

1. Yevics, P. Adapted, with permission, from *Disaster Planning: Protecting Your Firm, Your Clients and Your Family.* Law Office Management Maryland State Bar Association, Inc., 2001.
2. Berliner, N. "What doctors do in an emergency." *Pharmaceutical Representative* (2006).
3. Zinkewicz, P. "Business interruption insurance—death protection for a business." *Rough Notes* (Jul 2005): 1–3. Available at: http://findarticles.com/p/articles/mi_qa3615/is_200507/ai_n14775710, accessed January 29, 2008.

Chapter 2

Technological Disasters

Technology ... is a queer thing. It brings you great gifts with one hand, and it stabs you in the back with the other.

—C. P. Snow, *New York Times,* March 15, 1971

Technological failures are a catastrophe for a medical office, potentially shutting it down in a nanosecond. These failures are frequent and can cause as much or more havoc than a natural disaster. Computers have been very helpful in the clinical areas of medical diagnosis and treatment and are now beneficial for business aspects as well. We have thus become dependent on technology to make our offices efficient and effective. When a computer program crashes or becomes nonfunctional, many practices will come to a screeching halt. For example, if you are using electronic medical records and the server goes down or Internet access is lost, your ability to provide quality care without a chart or access to a patient's medical records makes continuity difficult and at times impossible, as you likely do not know each patient's past history, his or her medications, and the results of his or her previous laboratory results and X-ray studies.

Most doctor or office practices have lost data or access to a patient's chart. The reason for this scenario, which occurs far too often in every practice, is simple: Today's computers are much more powerful and capable than devices of just a few years ago; therefore, we depend heavily on them. Because they are increasingly complex, they can fail in new and distressing ways.[1]

Fortunately, most computer disasters can be prevented, and if they do occur, the down time can be shortened considerably by proper preparation. This chapter provides the necessary steps for preventing and recovering quickly from a computer disaster.

Key Questions

These questions must be asked in order to prevent a computer disaster:

1. Is the computer backed up at least once daily?
2. Is the data stored offsite so that data can be restored if necessary?
3. Have you conducted a test of the backup system on a regular basis to ensure that the restore programs are truly functional?
4. Do you use secure encryption techniques to prevent unauthorized people from having access to your computers?
5. Are the passwords unique? Can they be easily identified by unauthorized persons?
6. If all of the computers were damaged or destroyed, do you have a timely method of getting new equipment?
7. Do you have proper protection against damaging electrical surges?
8. Are you using state-of-the-art virus protection that prevents destruction or corruption of your data?

This chapter discusses possible corrective actions so that a technological disaster either does not occur or is short lived.

The most common problem for medical offices is the loss of data; therefore, a secure backup system is imperative. Be sure to back up the data on a regular basis—at least once a day. If the key to success in real estate is location, location, location, then the best advice regarding loss of data is back up, back up, back up! Other simple practices can provide a deeper peace of mind and can avoid devastating computer surprises.

Complete protection, however, requires recognizing that technological disasters come in many varieties and severities. Your modem or mouse can break, a program application might fail repeatedly, a critical data file may disappear, the Windows registry can be corrupted, or your hard drive can crash. Your annoyance, irritation, or panic level will depend on whether you are facing a nuisance problem or disaster.

Proper planning lessens the severity of many problems. Thus, first you must analyze what you use your computing equipment for, listing what you depend on (hardware, software, electricity, Internet connection, etc.) and deciding what you would do if various problems occur. Create scenarios of various technological disasters that might occur in your office. Play devil's advocate, and assume that the worst can happen. What would you do to fix them or get them functioning as soon as possible? Create a disaster plan for your computers, and then have regular drills to check that the plan is indeed functional. Some practices and small businesses are dependent on computers and thought that they had a backup plan in place; instead, however, they discovered that the backup program did not capture the data. The lost data could be restored only with expensive disc doctors. Clearing computers of viral infections requires both manpower and other resources. This translates into tens or even hundreds of thousands of dollars in desktop cleanup costs for each virus outbreak within a practice. If actual data destruction has occurred, the costs can be immense, possibly immeasurable; therefore, regular practice sessions to check the backup system are imperative. These procedures always take longer the first time; however, this practice often reveals false assumptions and missing steps. In Chapter 6, we cover the methods that are available to back up and retrieve data.

PROTECTING YOUR COMPUTER FROM UNWANTED EYES AND CYBERTHIEVES

An untimely computer crash can significantly impact your practice. These crashes can range from moderately frustrating application shutdowns and "not responding" pop-ups to the fearsome hard disk wipeout, bringing your entire computer system to a complete halt. Even the message "restart your computer" and hitting control, alt, delete will not help to jump start your computers. Possible troubles include viruses, spyware, adware, and unauthorized persons gaining access to your computer.[2]

Viruses

Perhaps the most virulent problem is a virus—a computer program that can copy itself, infect your hard drive without permission, and corrupt your data. A virus will spread from one computer to another in the office by transmission over your network or even carrying it on a removable medium such as a floppy disk, CD, USB drive, or more commonly, the Internet. Viruses are sometimes confused with computer worms and Trojan horses. A worm can spread itself to other computers without needing to be transferred as part of a host, and a Trojan horse is a file that appears harmless until executed.

Many personal computers are now connected to the Internet and to local area networks, facilitating the spread of these dangerous codes and viruses. Some viruses are programmed to damage the computer by corrupting programs, deleting files, or reformatting the hard disk. Others are not designed to do any damage but simply to replicate themselves and perhaps make their presence known by presenting text, video, or audio messages. Even these benign viruses can create problems for the computer user. They typically use large amounts of computer memory, thus causing legitimate programs to run very slow and disrupting the work processes within your office. In addition, many viruses are bug ridden, cause erratic behavior, and may lead to system crashes and data loss.

Adware and Spyware

Adware and spyware are also damaging and can impact the performance of your computer programs. Adware is software that inexplicably displays advertisements on your computer screen, even if employees are not using the Internet. Spyware is software that sends your personal information to a third party without your permission or knowledge. This can include information about websites that you or your employees visit or something more sensitive, like your user name and password. Unscrupulous companies often use these data to send you unsolicited targeted advertisements and could potentially hack into your computers to obtain protected health care data on your patients.

Security Software

Virus scanners, adware, and spyware programs work two ways. The more common method scans the file against all known viruses that are known each time the file is opened. The second method takes a blueprint of every file ahead of time and stores it in a database. The next time a file is opened, the software recomputes and checks the hard drive and compares it with the one in the previous database to see whether the file has changed. If it has changed, the program scans the file for viruses. If not, the file is considered virus free. Because most files are virus free, this method is faster because recomputing is considerably faster than comparing each file previously written to the database. Instead of scanning for all virus signatures each time a file is opened, the antivirus program performs a full scan on the file only one time. A full scan has to be performed only once in the beginning in order to determine whether the file is free of viruses. Each additional scan compares the new files with the older files already in the data base of checked files, making this method considerably faster.

Antivirus software that is noted for fast performance is available from Sophos (www.sophos.com); it runs on a wide variety of platforms, including Windows, Mac, Linux, Unix, NetWare, and Network Appliance storage systems.

The ideal security suite for small practices is one that supplies a broad array of effective protection for all of your computers, including laptops, desktops, and servers that are running Microsoft Windows and Mac OS X. The protection must guard against the entire spectrum of threats, including not only the known viruses, but also new variants of known viruses and potentially unwanted applications, including adware and spyware.

The ideal security system must also be easy to install and configure. Furthermore, the program must be simple to manage and monitor. Administrators and information technologists must have the ability to schedule scans and monitor protection status, and individual users should be able to identify virus threats with alerts in real time.

Finally, the ideal security suite unifies all of these capabilities with the smallest possible demands on the computer's processor and memory so as not to interfere with core business tasks.

Security software programs offer a great number of choices: single-function products as well as integrated ones and suites that have been designed for consumers, small practices, or large mega groups. Three well-known programs include Sophos, Symantec or Norton Utilities (www.symantec.com), and McAfee (www.mcafee.com); all three target small- to medium-sized medical practices. These programs contain protection against viruses, spyware, and day-zero threats.

Virus scanners are available according to a subscription model. The McAfee and Sophos products are backed by 24/7 support; for the Symantec product, such round-the-clock support is a premium option. The Sophos and Symantec suites allow you to install the management component on a single existing server on your network, whereas the McAfee-managed service hosts the management component and server. Two factors are essential for a computer security program that is targeted to small- to medium-sized medical practices. First, the suite must provide effective protection against an array of threats. Second, the suite must be easy to install and administer.

Although computer crashes, electrical surges, or data theft can wreck havoc with your computers, viruses, worms, and spyware are also high risk. Although the term *spyware* suggests software that secretly monitors the user's behavior, the functions of spyware extend well beyond simple monitoring. They can collect various types of personal information and can also interfere with user control of the computer in other ways, such as installing additional software, redirecting web browser activity, or diverting advertising revenue to a third party.

Each of the products that we evaluated is sold according to a subscription model. You pay for a minimum of five users and at least 1 year of periodic updates, thus ensuring that you are protected against emerging threats.

In response to the emergence of spyware, a small industry has sprung up dealing in antispyware software. Running antispyware software has become a widely recognized element for Microsoft Windows desktop computers.

The Federal Trade Commission has proposed defining spyware as "software that aids in gathering information about a person or organization without their knowledge and that may send such information to another entity without the consumer's consent, or that asserts control over a computer without the consumer's knowledge."[3] Infection by spyware can have a number of potentially harmful consequences and can result in computer users having to spend considerable time and money to resolve these problems. One potential effect is a dramatic slowing of the infected computer's performance, or worse, a system crash. Another potential consequence of spyware can be changes to the computer's settings, including security settings that can substantially reduce the consumer's ability to control his or her own computer.

Spyware is capable of more serious harm: the theft of personal information resulting from the spyware monitoring a computer user's activities. Furthermore, spyware can lead to security breaches of personal health information stored in computerized databases. Although there are legitimate uses for monitoring software, such as employers monitoring the computer use of employees and parents monitoring their children's online activities, cyberthieves are increasingly using these programs to facilitate identity theft.

CYBERTHIEVES

In May 2006, personal data on more than 26 million U.S. veterans fell into the hands of thieves. The data were on a laptop and external drive stolen in an apparent random burglary from the Montgomery County, Maryland home of a Department of Veterans Affairs computer analyst. The computer disk contained the names, social security numbers, and birth dates of every living veteran from 1975 to the present.[4] Although the theft did not include health records or financial information, the potential for loss of such sensitive data is a real possibility that every medical practice must guard against. Cyberthieves can use several different kinds of spyware to commit identity theft.

Keyloggers

Keylogger programs monitor the actions of a computer user in order to gain access to financial accounts on the web. They target specific information such as passwords, account numbers, and user names and can record every keystroke that a computer user makes, thereby capturing e-mail addresses and websites visited and any other information the user enters. A file containing this information is sent to the cyberthief, who can extract the computer user's sensitive personal information. One study estimates that nearly 10 million households in America have a computer infected with a keylogger program.[5] Researchers have noted that the number of websites spreading keylogger programs has increased dramatically in recent months (Figure 2-1).

Redirectors

These programs redirect web traffic to websites that the user did not intend to visit. For example, a computer user types in the name of his

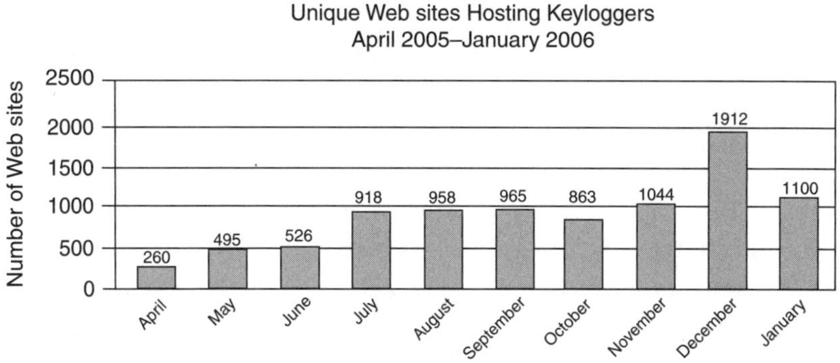

Figure 2-1 Number of Web sites hosting keyloggers.
Source: From www.antiphishing.org/reports/apwg_report_jan_2006.pdf. Anti-Phishing Working Group, *Phishing Activity Trend Report*, January 2006, courtesy of anti-phishing.org.

or her financial institution and is unknowingly redirected to a fraudulent website that mimics the legitimate Web site. The user is then tricked into providing password, user name, and account information.

Remote Access

This type of spyware allows cyberthieves to have remote access to and control of the user's computer system and can allow them to collect and share personal information without the user's knowledge. Remote access can also allow the user's computer to be used for illegal activities, such as sending spam or phishing e-mails or spreading keylogger programs.

RESPONSES TO SPYWARE

Authorities are currently pursuing a number of responses to the challenge of spyware.

Technological

Antispyware software, which identifies and removes spyware from a user's computer, is commercially available. Such antispyware programs typically can identify only known spyware programs, creating a lag time between the distribution of a spyware program and the ability of antispyware programs to detect it.

Behavioral Change

Computer users can change their online behavior, thus minimizing the risk of encountering spyware. This includes not opening e-mail attachments, not visiting particular websites, not downloading music or video files, or changing web browsers. A recent study indicates that computer users who have been victimized by spyware are more likely to make these types of behavioral changes than are individuals who have not.[6]

Legislative Change

State and federal legislators have addressed the problem of spyware. In 2005, antispyware legislation was introduced at the federal level and in 28 states. Such legislation typically seeks to establish criminal penalties for unauthorized dissemination and use of spyware programs.[7]

Although Internet users can reduce the possibility of being victimized by spyware, cyberthieves have substantial financial incentives to continue to develop newer and more sophisticated spyware programs. As a result, spyware programs will likely become more numerous, more difficult to detect, and more effective at gathering personal information; therefore, consumers need additional defenses against the identity theft that can result from spyware. One such defense includes the imposition of a security freeze on their credit report information to prevent cyberthieves from using stolen personal information to open fraudulent accounts.

PROTECTING AGAINST SPYWARE AND ADWARE

First, discourage your office staff from downloading anything from the Internet onto your office computer. Make sure that the programs you install onto your hard drive do not contain adware. Be aware that some free programs might include adware. If you are unsure, read the license agreement carefully (these are usually shown directly or through links as part of the installation process). Also, check the publisher's website. If you are still not sure, search Google Groups (http://groups.google.com/) for the name of the program and the keywords "adware" or "spyware." If nothing is mentioned, then you are probably okay.[8]

You can also prevent the unwanted installation by using a pop-up blocker. Spyware can install onto your hard drive after you click on a deceptive link. If a pop-up blocker is used on your computer, you will not even be tempted to click those links. Two of the best pop-up blockers are completely free: MSN Toolbar (http://toolbar.live.com/?mkt=en-us) and Google Toolbar (http://toolbar.google.com/T4/index_pack.html).[8]

PASSWORDS TO PROTECT YOUR COMPUTERS

Computer passwords are used to help identify and distinguish users, to verify their access on a computer or computer network, and to help restrict and prevent users from accessing confidential data or restricted programs. You can make your passwords more secure by the following rules:

- Do not use a recently used password.
- Change your password at least every 3 to 6 months.
- Create a password that is at least six characters long.
- Create a password with both digits and letters.
- Do not create a password with a family name or family pet.
- Do not create a password that is your phone number, house number, apartment number, and so forth.
- Create a password that is not in a dictionary.
- Create passwords with spaces in them.

Examples of some of the most commonly used passwords are listed here:

- no password (simply pressing enter)
- admin

- root
- password or PASSWORD
- god
- letmein
- love
- monkey
- pass
- sex
- 123, 1234, 12345, or 123456
- secret
- money
- asdf
- qwerty
- passme

Users also commonly use family and pets names such as Charlie, Thomas, or Fluffy; sports teams or sports players' names; phone numbers or parts of a phone number; and social security numbers or parts of a social security number.

Try using these rules to help secure your network and computers:

- Require that passwords be changed every 3 months (90 days). Almost all network operating systems have features that prompt users to change passwords after a specified time.
- Set a minimum password length. Most network operating systems support the ability to set a minimum password length.
- Set up a password history, if available. If the network operating system supports a password history, enable it to prevent the same password from being used.
- Enable account lockout threshold. This option disables an account after so many failed log-in attempts. Generally, three attempts with a duration of 60 minutes are sufficient.
- Do not write your password on the computer.

This is a listing of good examples of passwords:

iamthe1	2bornot2b	My1PASSword
1PaSsWoRd1	pa$$word	epoh

As illustrated previously, you can see passwords and famous quotes with numbers, passwords with numbers or special characters, or words spelled backwards, such as epoh (hope).

FIREWALLS

A firewall is a computer software system that is designed to prevent unauthorized access to or from your computers. Firewalls can be implemented in both hardware and software or in a combination of both. Firewalls are frequently used to prevent unauthorized Internet users from accessing private networks connected to the Internet, especially intranets, which is "a private version of the Internet," or as a version of the Internet confined to an organization. All messages entering or leaving the intranet pass through the firewall, which examines each message and blocks those that do not meet the specified security criteria. A firewall is simply a program or hardware device that filters the information coming through the Internet connection into your private computer system. If an incoming packet of information is flagged by the firewall filters, it is not allowed through.

Suppose your practice has between 10 and 100 employees and that you have dozens of computers connected via a server. The practice will therefore have network cards connecting them together. In addition, the practice will have one or more connections to the Internet through something such as a DSL or T1 or T3 lines. Without a firewall, all of those computers are directly accessible to anyone on the Internet. A knowledgeable person can hack or probe into those computers, make connections to them, and gain access to sensitive data. If one employee makes a mistake and leaves a security hole, hackers can get to the computer and exploit the hole.

With a firewall in place, the landscape is much different. A company will place a firewall at every connection to the Internet (e.g., at every T1 line coming into the company). The firewall can implement security rules. For example, one of the security rules inside the company might be this:

> Out of the dozens of computers inside this company, only one of them is permitted to receive public FTP (file transport

protocol) traffic. Allow FTP connections to only one computer, and prevent them on all others, making the network safer and less likely to be breached by hackers.

A practice can set up rules and can control how employees connect to websites. With a firewall, you can control whether files are allowed to leave the office over the network. A firewall establishes tremendous control over how people use the network.

Firewalls use one or more of three methods to control traffic flowing in and out of the network:

- Packet filtering. Packets (small chunks of data) are analyzed against a set of filters. Packets that make it through the filters are sent to the requesting system, and all others are discarded.
- Proxy service. Information from the Internet is retrieved by the firewall and is then sent to the requesting system and vice versa.
- Stateful inspection. This is a newer method that does not examine the contents of each packet but instead compares certain key parts of the packet to a database of trusted information. Information traveling from inside the firewall to the outside is monitored for specific defining characteristics, and then incoming information is compared with these characteristics. If the comparison yields a reasonable match, the information is allowed through; otherwise, it is discarded.

A firewall is considered a first line of defense in protecting private health information. For greater security, data can be encrypted.

ENCRYPTION

Medical offices must keep their data secure and not amenable to hackers and unauthorized persons. Also, doctors and staff are now using the e-mail to communicate with patients; therefore, the communication must be secure. Today, disaster planning has to include not only data protection but also data security. The use of encryption software is the most effective way to achieve data security. To read an encrypted file, you must have access to a secret key or password that enables you to decrypt it.

There are two main types of encryption: asymmetric encryption (also called public-key encryption) and symmetric encryption.

Asymmetric Encryption

A public key encryption system uses two keys: a public key, which is known to everyone, and a private or secret key, which is known to only the recipient of the message. When person A wants to send a secure message to person in practice B, he or she uses B's public key to encrypt the message. Person B then uses his or her private key to decrypt it.

Public and private keys are related in such a way that only the public key can be used to encrypt messages, and only the corresponding private key can be used to decrypt them. Moreover, it is virtually impossible to deduce the private key even if you know the public key.

Public-key systems, such as Pretty Good Privacy, are becoming popular for transmitting information via the Internet. They are extremely secure and relatively simple to use. The only difficulty with public-key systems is that you need to know the recipient's public key to encrypt a message for him or her.

Symmetric Encryption

Symmetic encryption is a type of encryption where the same key is used to encrypt and decrypt the message. This differs from asymmetric (or public-key) encryption, which uses one key to encrypt a message and another to decrypt the message. In symmetric-key encryption, each computer has a secret key (code) that it can use to encrypt a packet of information before it is sent over the network to another computer. Symmetric key requires that you know which computers will be talking to each other so that you can install the secret key code on each one. Symmetric-key encryption is essentially the same as a secret code that each of the two computers must know in order to decode the information. The code provides the key to decoding the message. Think of it like this: You create a coded message to send to a friend in which each letter is substituted with the letter that is two down from it in the alphabet. Thus, "A" becomes "C," and "B" becomes "D." You have already told a trusted friend that the code is "shift by 2." Your friend gets the message and decodes it. Anyone else who sees the message will see only nonsense. For more on how encryption works, see this website: http://computer.howstuffworks.com/encryption2.htm.

SURGE PROTECTION

When building a computer system, you will probably want to buy a surge protector.[9] Most designs serve one immediately obvious function: They let you plug multiple components into one power outlet. Thus, this is definitely a useful device. The main purpose of a surge protector system is to protect electronic devices from electrical surges.

A power surge, or transient voltage, is an increase in voltage significantly above the designated level in a flow of electricity. In normal household and office wiring in the United States, the standard voltage is 120 volts. If the voltage rises above 120 volts, a surge protector prevents the increased voltage from destroying your computer.

There are two types of increased voltage: a surge and a spike. An increase in voltage lasts 3 or more nanoseconds (billionths of a second) is called a surge. A spike lasts for only 1 or 2 nanoseconds.

If the surge or spike is high enough, it can inflict some heavy damage on your computers. Even if increased voltage does not immediately break your machine, it may put extra strain on the components, thus shortening the computer's life expectancy.

LEVELS OF SURGE PROTECTION

All surge protectors are not created equal. In fact, there is a tremendous range in both performance and price of protection systems. At one end is the basic $5 surge protector power strip, which offers very little protection; on the other end, there are systems costing hundreds of dollars, which protects against pretty much everything short of lightning striking nearby. Picking out a protector system is a matter of balancing the cost of the system with the cost of losing data or electronic equipment.

To protect your equipment from surges, you need individual surge protectors for each outlet. These power strips range a great deal in quality and capacity. There are three basic levels of power strip surge protectors:

- Basic power strips. These are basic extension cord units with five or six outlets. Generally, these models provide only basic protection.
- Better power strips. For $15 to $25, you can get power strip surge protectors with better ratings and extra features.
- Surge stations. These large surge protectors fit under your computer or on the floor. They offer superior voltage protection and advanced line conditioning. Most models also have an input for a phone line to protect your modem from power surges and may feature built-in circuit breakers. You can get one of these units for as little $30, or you can spend more than $100 for a more advanced model.
- Uninterruptable power supplies (UPSs). Some units combine surge protection with a continuous UPS. The basic design of a continuous UPS is to convert AC power to DC power and to store it on a battery. The UPS then converts the battery's DC power back to AC power and runs it to the AC outlets for your electronics. If the power goes out, your computer will continue to run, feeding off the stored battery power. This will give you a few minutes to save your work and shut down your computer. The conversion process also gets rid of most of the line noise coming from the AC outlet. These units tend to cost $150 or more.

An ordinary UPS will give you a high level of protection, but you should still use a surge protector. A UPS will stop most surges from reaching your computer but will probably suffer severe damage itself.

Shopping for a surge protector is tricky because a lot of nearly worthless products are on the market. Research into a particular model is the best way to ensure good results, but you can get a good idea of a product's performance level by looking for a few signs of quality. First, look at the price. Do not expect much from any surge protector that costs less than $10. These units typically will not protect your system from bigger surges or spikes. Of course, high price does not promise quality. To find out about the unit's capability, check out its Underwriters Laboratories (UL) ratings. UL is an independent, not-for-profit company that tests electric and electronic products for safety. If a protector does not have a UL listing, it is probably ineffective. Be sure that the surge protector is listed as a transient-voltage surge suppressor. This means that it meets the criteria for UL 1449, UL's minimum performance standard for surge suppressors. Better surge protectors may come with some sort of guarantee of their performance. Consider a protector that comes with a guarantee on your computer. If the unit fails to protect your computer from a power surge, the company will actually replace your computer. This is not total insurance, of course, as you will still lose all of the data on your hard drive, thus costing you plenty; nevertheless, it is a good indication of the manufacturer's confidence in its product.

DATA REMOVAL AND ERASING HARD DRIVES

Many practices will discard a computer, giving it to a school or simply throwing it away after attempting to remove the data from the hard disk. Unless the hard drive is wiped clean using special programs, the data can be recovered by unauthorized persons. It is almost impossible to remove all of the information on your hard drive with the structure and functionality of the operative systems currently available. Format and delete commands do not remove information stored on your computer. These commands only alter the structure of the drive, leaving most of the data intact and recoverable with available software tools. Therefore, make sure that hard drives are erased beyond any reasonable ability to recover data. Deleting files and formatting your hard drives are not adequate methods of permanently removing data. Either hire an information technology professional to remove the data or use a data-removal utility, which will erase the data and replace it with 1s and 0s. Whether your data are sent to the recycle bin or your entire drive is formatted and repartitioned, the chance of unauthorized discovery is very real and poses issues of risk and liability. Securely wipe hard drives and overwrite, and delete and destroy privileged data with data-cleansing programs. Data-removal utilities such as CiberCide (http://www.cyberscrub.com/cybercide/) and DataEaser (http://www.ibas.com/data-erasure) are HIPAA compliant.

IMPACT OF FIRE ON DATA LOSS

According to the National Fire Protection Association, there were 125,000 nonresidential fires in 2001, with a total of 3.231 billion dollars in losses.[10] Industry studies tell us that 43% of the businesses that are closed by a fire never reopen; 29% of those that do open fail within 3 years. Thus, when designing data centers, fire prevention, detection, and suppression are always top concerns. Fires in data centers are typically caused by power problems in raceways, raised floors, and other concealed areas. A fire that destroys data in your office can be catastrophic to your practice. A fire that occurs in the data center can be caused by arson, corporate sabotage, and natural occurrences such as lightning and power surges. Fire prevention provides more protection against fire than any type of fire detection or suppression equipment available. If an environment is incapable of breeding a fire, then there will be no threat of fire damage to the facility. If a fire does occur, the next step is to detect it. Before fire alarms were invented, watchmen were responsible for spotting fires and alerting others. Now there are a number of advanced detectors that can detect fire in its earliest stages and then notify a center that notifies personnel and suppression systems. There are many ways of detecting and suppressing fires, but only a few are recommended for data center applications. In a data center, the main goal of the fire protection system is to get the fire under control without disrupting the flow of business and without threatening the personnel inside.[11]

Fires associated with data centers include fires involving ordinary combustible materials such as paper, wood, cloth, and some plastics and fires involving live electrical equipment. The principle of fire prevention and detection is based on the "Fire Triangle," which consists of oxygen, heat, and fuel. All three conditions must all interact for a fire to take place. Fires can be extinguished if one or more of these three elements are taken away; therefore, fire extinguishing methods can vary depending on which element(s) is removed. For instance, carbon dioxide systems reduce oxygen by displacing it with a gas that is heavier and colder than oxygen. Because carbon dioxide is much colder than the fire, it hampers its progression by taking away heat.

After a fire is started, it is often categorized in stages of combustion. There are four stages of combustion: (1) the incipient stage or precombustion, (2) visible smoke stage, (3) flaming fire stage, and (4) intense heat stage. As a fire progresses through these stages, many factors increase exponentially, including smoke, heat, and property damage—not to mention risk of life, which becomes critical as smoke density increases. Fire research has shown that the incipient stage allows for the largest window of time to detect and control the progression of a fire. In this window of time, fire detection systems can mean the difference between availability and unavailability and whether your data are saved or damaged beyond retrieval. The longer the fire burns the more products of combustion, which then leads to a higher chance of equipment failure even if the fire is successfully extinguished. These products of combustion may be electrically conductive and can also corrode the circuits on computers.

CHOOSING A FIRE PROTECTION SOLUTION

For the purposes of designing a fire protection solution for a data center, three conditions should be met: (1) identify the presence of a fire, (2) communicate the existence of that fire to the occupants and proper authorities, and (3) finally contain the fire and extinguish it if possible. Being familiar with all technologies associated with fire detection, alarming, and suppression will ensure a sound fire protection solution. Of course, before selecting detection and suppression systems, computer advisors must assess and evaluate the potential hazards and issues regarding preservation of data. For example, will the data center have raised floors? Will it have high ceilings? Will personnel occupy the area? Will detectors be obstructed in any way? These questions should be answered before the proper fire protection solution is chosen.

FIRE DETECTION SYSTEM TYPES

Three main types of detectors are available: smoke detectors, heat detectors, and flame detectors. For the purposes of protecting a data center, smoke detectors are the most effective. Heat and flame detectors should not be used in proximity data centers, as they do not provide early warning for the protection of your valuable data assets.

Spot-Type Smoke Detection

Spot-type or conventional smoke detectors can cover an area of about 900 square feet. Most data centers and computer rooms require additional space to compensate for the high air flow required in these environments to prevent overheating of the computers. A default standard for high air movement, data-sensitive areas is usually one detection device per every 250 feet.[2] Spot-type detectors are effective in small data centers and computer rooms. Although more expensive, intelligent detectors are available, they would add little value in these smaller spaces. Two types of spot-type detectors are available: photoelectric and ionization.

Photoelectric detectors work by using a light source or beam and a light sensor perpendicular to it. When nothing is in the chamber, the light sensor does not react. When smoke enters the chamber, some of the light is diffused and reflected into the light sensor, causing the alarm to sound.

Ionization detectors use an ionization chamber and a small amount of radiation to detect smoke. Normally, the air in the chamber is being ionized by the radiation, causing a constant flow of current that is monitored by the detector. When smoke enters the chamber, it neutralizes or disrupts the ionized air, thereby causing the current to drop. This triggers the detector into an alarmed condition.

Intelligent Spot-Type Smoke Detection

Intelligent spot-type detectors are very similar to conventional spot-type detectors; however, they report more precisely the location of a fire. These detectors are intelligent because they are able to send information to a central control station, thereby pinpointing the exact location of the smoke. Some intelligent spot detectors have the ability to compensate automatically for changing environments such as humidity and dirt accumulation. They can also be programmed to be more sensitive during certain times of the day; for instance, when workers leave the area, sensitivity will increase. Intelligent spot-type detectors are commonly placed below raised floors, on ceilings, and above drop-down ceilings; however, modified spot detectors are also used in air handling ducts to detect possible fires within the HVAC (heating ventilation air conditioning) system. By placing detectors near the exhaust and the intake of CRAC units (computer room air conditioners), detection can be accelerated.

Air-Sampling Smoke Detection

Air-sampling smoke detection, sometimes referred to as a "very early smoke detection" (VESD) system, is usually described as a high-powered photoelectric detector. Air-sampling systems use an advanced detection method using a very sensitive laser, much more powerful than the one contained in a common photoelectric detector. As the particles pass through the detector, the laser beam is able to distinguish them as dust or by-products of combustion. An air-sampling system is comprised of a network of pipes attached to a single detector, which continually draws air in and samples it. The pipes are typically made of PVC but can also be copper. Depending on the space being protected and the configuration of multiple sensors, these systems can cover an area of 2,500 to 80,000 square feet, which would be more than ample for even the largest medical practices and their computer data.

Smoke detection is dependent on three variables: (1) the sensitivity of the detector, (2) the clarity of the smoke path leading to the detector, and (3) the density of the smoke after it reaches the detector. In an area such as a data center, where the air flows are rapid, it becomes difficult to detect smoke with a spot-type detector, especially in the incipient stage of a fire. This makes VESD an ideal smoke detection solution for high-availability data centers. The air-sampling system is designed to detect the particles of combustion such as those released from PVC wire during the initial stages of heat buildup. When the smoke particles drift through the pipes and into the detector, a photo detector or a laser beam differentiates the particle as dust or as a by-product of combustion. This detection process can be up to 1,000 times more sensitive than a photoelectric or ionization smoke detector.

Linear Thermal Detection

Linear thermal detection is a method of detecting hot spots in cable trays or cable runs. Generally, it is not used in enclosed and air-conditioned computer rooms or data centers. Linear thermal detection is composed of at least two heat-dependent conductors. When a set temperature is reached, the two conductors cause an alarm condition that is detected at the main control panel. The control panel can then notify personnel and pinpoint the location of the hot spot.

FIRE-SUPPRESSION SYSTEM TYPES

After a fire is detected in a data center, it must be quickly extinguished with no effect on the data center operation. Regardless of the method employed, a means should be available to abort the suppression system in the event of a false alarm, as water, carbon dioxide, or other chemicals should not unnecessarily damaging computers or sensitive data.

Foam

Foam, formally called aqueous film-forming foams, is generally used in liquid fires. Because foam is electrically conductive, it should not be used in data centers.

Dry Chemical

Dry chemical or dry-powder systems can be used on a wide variety of fires and pose little threat to the environment. Different types of powders can be used depending on the type of fire. They are electrically nonconductive but are not recommended for data centers because of the residue that remains after discharge.

Water Sprinkler Systems

Water sprinkler systems are designed specifically for protecting the structure of the building. They are discharged when the valve fuse opens. Normally, water sprinklers are not recommended for data centers; however, depending on local fire codes, they may be required. In this case, a preaction system would be recommended. Installing a sprinkler system during construction can range from $1 to $2 per foot,[2] whereas retrofitting an existing building costs $2 to $3 per foot.[2]

Water Mist Systems

Water mist systems discharge very fine droplets of water onto a fire. Because of the small droplet size, there is a dramatic decrease in water consumption. Also, because mist systems use less water than conventional sprinkler systems, they require less storage space. These systems are extremely safe and pose no threat to the environment. This fine mist of water extinguishes the fire by first absorbing heat from the fire, thus causing a vapor and a barrier between the flame and the oxygen needed to sustain it. This change of state (liquid to gas) makes this water mist system very effective.

Water mist systems are gaining popularity because of their effectiveness; however, evidence suggests that equipment failure and data loss can result from a discharge caused by the high level of humidity introduced into the data center.

Fire Extinguishers

Sometimes the oldest method of fire suppression is the best. Fire extinguishers these days are essentially the same as they have always been—they are easy to use and can be operated by nearly anyone. Fire extinguishers are valuable to data centers because they can extinguish a fire before the main suppression system discharges. Various types of fire extinguishers have been approved for use in data centers. HFC-236fa (or more commonly called by its trade name, FE-36) can be used in occupied areas. They are environmentally safe and leave no residue on discharge because they are discharged as a gas. These clean agents extinguish fires by removing heat and chemically preventing combustion.

Total Flooding Fire Extinguishing Systems

Total flooding fire extinguishing systems, sometimes referred to as clean agent fire-suppression systems, can be used to suppress fires in data critical areas. A gaseous agent flooding fire-suppression system is highly effective in a well-sealed, confined area, which makes a data center an ideal environment. It typically takes less than 10 seconds for an agent to discharge and fill the room. The agent is contained in pressurized tanks, and the number of tanks used depends on the total volume of the room being protected, as well as the type of agent used. The hidden areas in a data center present the biggest threat of fire. If wires are cracked, damaged, loose, or otherwise poorly maintained in an open area, a routine visual inspection should uncover the problem, and repairs can be made immediately. Discovering a problem in a closed area is far more difficult. Unlike water-suppression systems, gaseous agents infiltrate even the hardest to reach areas, such as inside equipment cabinets. Later, the gas and its by-products can be vented out of the data center with very little environmental impact and no residue.

These agents are nonconductive and noncorrosive, and some can safely be discharged in areas that are occupied by your employees. The name "clean agents" is commonly used because they leave no residue and cause no collateral damage. For years, Halon has been used as the agent of choice; however, it was phased out in commercial applications because of its ozone-depletion properties.

Gaseous Agents

A fire-extinguishing agent is a gaseous chemical compound that extinguishes a fire by means of "suffocation" and/or heat removal. Given a closed, well-sealed room, gaseous agents are very effective at extinguishing fires and leave no residue. In the 1960s, Halon 1301 was widely used throughout various industries, given its effectiveness in fighting fires. Currently, Halons are banned because they are thought to deplete the ozone.

Gaseous agents are divided into two categories: inert gases and fluorine-based compounds. Inert gases sold under the trade names of Pro-Inert and Inergen are the most widely accepted and commercially available today. Carbon dioxide is an inert gas, which reduces the concentration of oxygen needed to sustain a fire by means of physical displacement. Because carbon dioxide is heavier than oxygen, it settles to the base of the fire and quickly suffocates it. That makes this type of agent unsafe for discharge in occupied areas and is therefore not recommended for data centers. If a carbon dioxide system is used in occupied areas because no suitable alternative is available, a proper evacuation plan should be in place, and safety mechanisms should be used to notify personnel to evacuate the effected areas before a discharge. A safety mechanism would be one that provides audible and visual queues to data center occupants 30 to 60 seconds before discharge. Another disadvantage of using carbon dioxide is the large number of storage containers that are required for effective discharge. This is obviously a poor choice for any data center where floor space is highly valued.

Inert agents are nonconductive, leave no residue, and are safe to discharge in occupied areas. They are stored as a gas in high-pressure tanks that can be located up to 300 feet away from the protected space. This is convenient considering that inert agents require a storage volume 10 times that of other alternatives available today, which would take up precious data center space. Inert agents are used in data centers, telecommunications offices, and various other critical applications.

Fluorine-Based Compounds

Although other alternative agents are approved by the National Fire Protection Agency, fluorine-based compounds are the most widely accepted and commercially available for the protection of high-value assets.

Fluorine has a zero ozone-depletion potential and an extremely low global warming potential. Fluorine can be stored as a liquid and is colorless and nearly odorless. Although at room temperature it is a liquid, it is discharged as an electrically nonconductive gas that leaves no residue and will not harm occupants; however, like in any other fire situation, all occupants should evacuate the area as soon as an alarm sounds. Fluorine extinguishes a fire by removing heat faster than it is generated and is discharged in 10 seconds or less.

PULL STATIONS

Pull stations, which allow a building occupant to notify everyone in the building of a fire, should be placed at every exit to the protected space. Once pulled, they can notify the fire department of the alarm. Pull stations are sometimes the best way to catch a fire in its incipient stage. No matter how sensitive a smoke detector may be, it is still no substitute for the human nose. A person can pick up the scent of smoke much earlier than any smoke detector can.

SIGNALING DEVICES

Signaling devices are activated after a pull station or a detector enters an alarm condition. Signaling devices provide audible and/or visual queues to building occupants as a signal to evacuate the building. Audible sounds may include horns, bells, and sirens and may be heard in various patterns. Visual signaling devices are crucial to notifying occupants who are hearing impaired. Strobes usually incorporate a Xenon flash tube that is protected by a clear protective plastic. They are designed with different light intensities measured in candela units. The minimum flash frequency for these strobes should be once per second.

CONTROL SYSTEMS

Fire-suppression and fire-detection products in a building are useless without a control system, commonly known as the fire alarm control panel. Control systems are the "brains" behind the building's fire-protection network. Every system discussed thus far is accounted for by the fire alarm control system. Fire alarm panels are either conventional panels or intelligent addressable panels working with detectors of the same type (conventional or intelligent/addressable) and with the same communication protocol. Depending on the panel, it can control the sensitivity levels of various components such as smoke detectors and can be programmed to alarm only after a certain sequence of events have taken place. The computer programs used by these systems allow a user to set certain time delays, thresholds, passwords, and other features. Reports can be generated from most intelligent panels, which can lead to improved performance of the fire protection system, for example, by identifying faulty sensors. After a detector, pull station, or sensor is activated, the control system automatically sets in motion a list of rules that have been programmed to take place. It can also provide valuable information to authorities.

All fire alarm control panels used in a suppression environment should be listed by UL for "releasing." This approval guarantees that the control panel incorporates the necessary protocol and logic to activate and control a fire-suppression system.

Raised Floors

Raised floors bring up some important issues with regard to fire protection in mission-critical facilities. Raised floor tiles conceal power and data cables as well as other combustible materials, such as paper and debris; therefore, all cabling should be placed overhead, where it is visible and can be easily inspected. Redundant air-sampling smoke-detector systems should be placed beneath as well as above the raised floor.

PROTECTING MISSION-CRITICAL FACILITIES

Now that all the fire protection components have been described, the last step is to bring them together to design a robust and highly available data center solution. Although various types of detection, suppression, and gaseous agents were described, not all of them are recommended for a highly available data center. The following is list of components that protect a data center with a goal of 24/7 coverage.

- Conventional spot-type detection
- Intelligent spot-type detection
- Air-sampling smoke detection
- Fire extinguishers
- Total flooding fire-extinguishing system
- Halon alternative clean agent
- Pull stations
- Signaling devices
- Control system/fire alarm control panel
- Fire control panel

INDUSTRY BEST PRACTICES

The following is a list of recommended practices for increasing the availability of a data center with respect to fire protection.

- Ensure that the data center is built far from any other areas of the practice that may pose a fire threat to the data center, such as a laboratory that has flammable chemicals.
- Emergency procedures should be posted on all fire alarm control panels.
- A smoke-purging system must be installed in the data center.
- All electrical panels must be free of any obstructions.
- All fire alarm pull stations should be consistently labeled to avoid any confusion.
- All fire extinguisher locations should be clearly identified and should provide information on what kind of fire to use it on.
- Any openings in the data center walls should be sealed with an approved fire-proof sealant.
- Each data center exit should have a list of emergency phone numbers clearly posted.
- Enforce a strict no-smoking policy in data control rooms.
- Equip the data room with fire extinguishers.
- Fire dampers should be installed in all air ducts within the data center.
- Fire protection systems should be designed with maintainability in mind. Replacement parts and supplies should be stored on site.
- Get approval from the fire marshal to continue operating the CRAC units when the fire system is in the alarmed state.
- If a facility is still using dry chemical extinguishers, ensure that the computer room extinguishers are replaced with a Halon alternative.
- Preaction sprinklers should be placed in the data center, as well as in the hallways.
- Provide a secondary water source for fire sprinklers.
- Sprinkler heads should be recessed into the ceiling to prevent accidental discharge.
- The fire-suppression system should have a secondary suppression agent supply.
- The data center should not contain any trash receptacles.

- All office furniture in the data center must be constructed of metal. Chairs may have seat cushions.
- Tape libraries and record storage within the data center should be protected by an extinguishing system. They should be stored in a fire-safe vault with a fire rating of more than 1 hour.
- Any essential supplies such as paper, disks, and wire ties should be kept in completely enclosed metal cabinets.
- Extension cords used to connect computer equipment to branch circuits should not exceed 15 feet in length.
- The use of acoustical materials such as foam and fabric used to absorb sound is not recommended in a data center.
- The sprinkler system should be controlled from a different valve than the one used by the rest of the building.
- All data center personnel should be thoroughly trained on all fire-detection and extinguishing systems throughout the data center. This training should be given on a regular basis.
- Air ducts from other parts of the building should never pass through the data center. If this is not possible, then fire dampers must be used to prevent fire from spreading to the data center.
- Water pipes from other parts of the building should never pass through the data center.
- Duct coverings and insulation should have flame spread ratings less than 25 and a smoke developed rating less than 50.
- Air filters in the CRAC units should have a class 1 rating.
- Transformers located in the data center should be a dry type or should be filled with noncombustible dielectric.
- No extension or power cords should be run under equipment, mats, or other coverings.
- All cables passing through the raised floor should be protected against chaffing by installing edge trim around all openings.
- Computer areas should be separated from other rooms in the building by fire-resistant–rated construction extending from structural floor slab to structural floor above (or roof).
- Avoid locating computer rooms adjacent to areas where hazardous processes take place.

COMMON MISTAKES

Some common mistakes made with regard to fire protection systems in a data center environment include the following:

- Having the fire system automatically shut down the CRAC unit. This will cause the computer equipment to overheat, resulting in down time.
- Using dry chemical suppression agents to extinguish computer room fires, as this will damage equipment. Dry chemical agents are very effective against fires but should not be used in a data center.
- Storing combustible materials underneath a data center raised floor.

Most fires in mission-critical facilities can be prevented if common mistakes are avoided and fire detection is properly specified and monitored. Human error plays a large role in preventing fire hazards and must be eliminated through training and procedures that are enforced.

BOTTOM LINE

Although disasters such as hurricanes, floods, and earthquakes get the most attention, especially in the media, technological disasters are the most common and the most likely to impact your practice. Technology is not foolproof, and even the most reliable power and Internet service provider connections can fail. More and more practices are becoming dependent on technology, computers, and access to the Internet; therefore, we must build in safety and use methods of protecting that valuable data. We depend on our computers and Internet access; thus, it is important to install an uninterruptible power supply and arrange backup Internet service. It is also imperative to regularly update and test your data protection against computer hazards.

After protecting your practice against these computer hazards, make a calendar entry every 4 to 6 months to review, update, and test your plan. Check your insurance coverage for your computers and other technologies. Disaster planning is not an event—it is a process that never ends.

REFERENCES

1. Goldberg, G. "Anticipating and Avoiding Personal Computer Disasters." Available at: http://www.aarp.org/learntech/computers/howto/a2004-10-21-disasters.html, accessed January 29, 2008.
2. Berger, S. "Handling Spyware." Available at: http://www.aarp.org/learntech/computers/howto/handling_spyware.html, accessed January 29, 2008.
3. Federal Trade Commission. Public Workshop: Monitoring Software on Your PC: Spyware, Adware, and Other Software. 69 *Federal Register* 8538 (February 24, 2004). Available at: http://www.aarp.org/research/technology/onlineprivacy/fs126_spyware.html.
4. Frieden, T., King, J., Walton, M. "Theft of vets' data kept secret for 19 days." Available at: http://www.cnn.com/2006/US/05/23/vets.data/, accessed January 29, 2008.
5. Krebs, B. "Hacking Made Easy," *Washington Post*, March 16, 2006.
6. Fox, S. "Spyware: The Threat of Unwanted Software Programs Is Changing the Way People Use the Internet." Pew Internet & American Life Project (July 6, 2005). Available at: http://www.pewinternet.org/pdfs/PIP_Spyware_Report_July_05.pdf, accessed January 29, 2008.
7. National Conference of State Legislators. "2005 State Legislation Relating to Internet Spyware or Adware." Available at: http://www.ncsl.org, accessed March 15, 2006.
8. Honeycutt, J. "How to Protect Your Computer from Spyware and Adware." Available at: http://www.microsoft.com/windowsxp/using/security/expert/honeycutt_spyware.mspx.
9. Harris, T. "How Surge Protector Work." Available at: http://www.howstuffworks.com/surge-protector.htm, accessed January 29, 2008.
10. National Fire Protection Association NFPA, Fire Analysis and Research Division. Information included from http://www.ptsdcs.com/whitepapers/62.pdf.
11. Avelar, V. "Mitigating Fire Risks in Mission Critical Facilities, White Paper #83." Available at: http://www.apcmedia.com/salestools/SADE-5TNRPF_R1_EN.pdf, accessed January 29, 2008.

ACKNOWLEDGMENT

Part of this chapter was provided courtesy of Victor Avelar and APC at http://www.apcmedia.com/salestools/SADE-5TNRPF_R1_EN.pdf.

Chapter 3

Preparing a Disaster Plan

Luck occurs when preparation meets persistence.

—Gary Player, professional golfer

Dr. Ron Kellum is a primary care doctor who practiced in Diamondhead, Mississippi, which is located on the Mississippi Gulf Coast. He practiced in a wooden building just 600 yards from the Gulf of Mexico. Before August 30, 2005, when Hurricane Katrina devastated the Gulf Coast area, he had made very little preparation and did not have a disaster plan in place. The hurricane caught him by surprise. Federal supplies did not come for 6 weeks, and he was trying to provide emergency care for a community of 10,000 people. Generous physicians throughout the region and pharmaceutical companies sent him supplies, which kept the practice functioning; thus, he was able to provide greatly needed emergency care to the community within 48 hours after the storm. Dr. Kellum advises all physicians to have a disaster plan in place, as they may not be able to count on state or federal assistance for days or weeks after a disaster.

Key Questions

1. Do you have a disaster plan in place?

2. If your building was flooded or had a fire, could you get your practice back in operation in 24 to 48 hours, or would it take you weeks or months to become operational?

3. Do you know how to contact all of your staff members if the phone lines were down and the usual methods of communication were not available?

4. Have you appointed a disaster plan captain to oversee recovery effort?

5. Would you be able to have access to your patients' charts after a disaster if you could not enter your building or practice area?

PREPARATION OF A DISASTER PLAN

This chapter features a detailed process for creating an effective disaster plan. It includes 13 important checklists that should be completed before the disaster occurs (e.g., creation of a first-aid kit and a phone tree to contact all personnel in the office and development of a key contact list for all of the necessary contacts for the ongoing administration of the office practice). The chapter also lists the necessary ingredients of a basic disaster supply kit. After reading this chapter, you will have the tools and necessary ingredients to create a disaster plan for your practice.

Consider this scenario: You have a practice that enjoys a prestigious reputation within the community. You have more than enough patients. Your facility is well located and is in prime condition. Now imagine that all of this investment is devastated by a disaster. This scenario is common and affects far too many medical practices. A loss of this magnitude can be mitigated, or at least minimized, if the practice has a carefully constructed disaster plan in place. By following the recommendations in this chapter, you can improve your chances of preserving your practice if a disaster happens. This chapter explains the steps physicians and office managers need to think about, to prepare, and to implement in order to prepare a disaster recovery plan so that operations can be resumed in a timely fashion.

Nearly every medical practice has a mentality of "if it ain't broke, don't fix it!" That philosophy may apply to your automobile, but not to your medical practice. It is also important to know that a Health Insurance Portability and Accountability Act (HIPAA) requirement (which states that health care organizations, including medical practices, must implement a disaster recovery process) makes good business sense.[1] The goal of the disaster plan is to have contingency plans in place to preserve the practice so that it can continue to offer its services to patients. In order for a plan to be useful, it must be created *before* a disaster or an interruption occurs. A carefully crafted disaster plan will ensure practice continuity and a prompt disaster recovery. Failure to have this in place can result in lost revenue and in patients going to practices that have a plan and that are functioning with as little down time as possible. The creation of a disaster plan essentially allows a physician to provide continuity of care and to continue generating an income stream, thus maintaining the financial viability of his or her practice.

Whatever one chooses to call it—disaster planning, emergency preparedness, or business continuity—the end result is always the same: getting your practice functioning in the event of an interruption. The problem causing the interruption could be one computer crashing or an entire network that is rendered nonfunctional. It could also be an electrical outage or the result of a terrorist activity. None of us can foresee the future, but we all have an opportunity to have some contingency plans in place in case any of these problems make our practices inoperable.

Part of writing a disaster plan is to imagine the possibilities of what can go wrong and to prepare contingency plans; however, you cannot possibly plan for every scenario; this would take an inordinate amount of time, and the plan would never be completed. The goal is not to create a separate plan that addresses every conceivable risk, but to produce one plan that addresses the most common risks or those most likely to occur in your geographic area. In other words, do not create one plan for a tornado, one for a flood, one for a blackout, and one for a power outage that disrupts your computers. Have just one plan that addresses the most likely scenarios or risk factors.

Before initiating your disaster plan, think about the impact on your patients if a disaster forces your practice to close or change the way you normally practice medicine. Will your patients be able to contact you if the phone system is inoperative? Where will you care for your patients if the hospital is closed? How will you access the patients' records if you do not have access to the charts? These questions will need to be considered and answered when preparing your disaster plan.

When analyzing risks, consider these factors:

- Historical: What types of emergencies have occurred in the community, at your facility, or in the region (e.g., fire, natural disasters, accidents, and utility outages)?
- Geographic: What can happen as a result of your location (e.g., proximity to flood-prone areas; hazardous material production, storage, or use; major transportation routes; and power plants)?
- Human error: What emergencies might be caused by employees? Are employees trained to work safely? Do they know what to do in an emergency? Human errors often result from poor training and supervision, carelessness, misconduct, substance abuse, and fatigue.
- Physical: What types of emergencies could result from the design or construction of the facility? Does the physical facility enhance safety? Consider the physical construction of the office, the facilities for storing combustibles or toxins, hazardous processes or byproducts, lighting, evacuation routes and exits, and shelter areas.

What could happen as a result of a computer crash, prohibited access to your office, loss of electricity, ruptured gas mains, water damage from flooding or ruptured water pipes, smoke damage, structural damage, air or water contamination, building collapse, trapped persons, or a chemical release? Ultimately, only four different scenarios need to be planned for, regardless of the catastrophe or interruption:

1. Only your office within the building is unusable. For example, one or more of the offices in your building become temporarily unusable because of a malfunctioning water sprinkler. In this situation, some of your contents and materials may be recoverable.
2. The entire building is gone. For example, a fire destroys the entire structure and all of its contents.
3. A temporary disruption of services, such as an electrical outage, causes your computers and Internet to be nonfunctional.
4. An impact in the region occurs, rendering the area uninhabitable for an unknown amount of time.

First, each practice must determine the most likely disasters that may impact the practice and then make preparations for each of these situations. Certainly, a wild fire that would potentially impact a practice in the Rocky Mountains will require different preparation than a hurricane that may hit the Gulf Coast area. Figure 3-1 illuminates the natural disasters that are more likely to occur in your area. Nearly all practices, however, are at risk for a computer crash, theft, and loss of electronic data; therefore, they must be prepared for such disasters. Begin by taking the following self-assessment (Table 3-1), to see how well prepared your practice is for a major disaster that might interrupt your practice.

Start creating your business continuity plan by completing nine key forms, which can be downloaded and printed from the CD-ROM included with this book. This vital information will help you to recover your essential business functions and help your employees to know what their responsibilities are. The property protection plan is a checklist that focuses on the natural hazards of wind, flood, earthquake, freezing weather, and wildfire. It covers your building and surroundings and is helpful whether you own or lease your building.

Next, prepare a disaster box (see Disaster Preparedness: Continuity of Operations Planning for Nonprofits, Louisiana Association of Nonprofit Organizations, www.lano.org).

48 | **Chapter 3:** Preparing a Disaster Plan

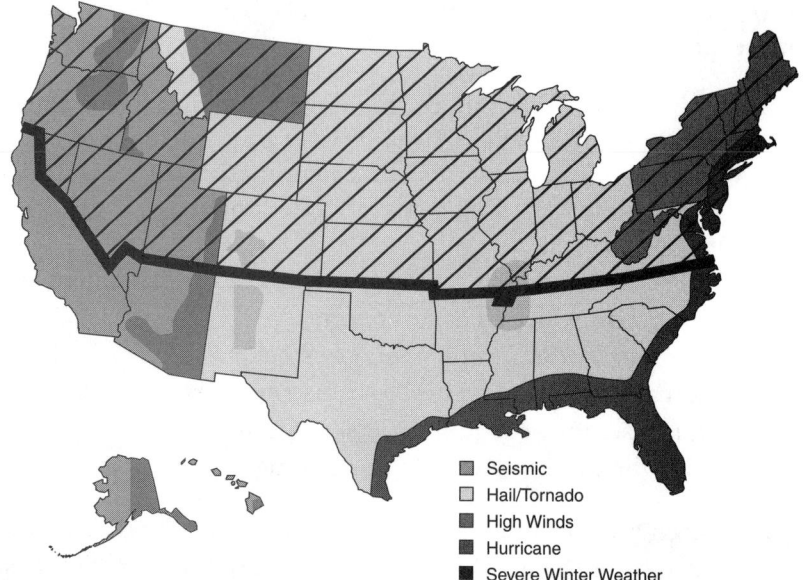

Figure 3-1 Risk factors for a natural disaster.

The disaster box is an essential component for every disaster plan. It contains all of your practice's core information. If your building or practice is unreachable because of a natural disaster or by flood, fire, or wind, the contents of this disaster box will help it to become functional in a timely fashion. The necessary contents of this disaster box are shown in Figure 3-2.

The practice disaster box should be stored somewhere outside of your practice (e.g., in the closet of one the physician's home or the practice manager). The box should be elevated off the floor in case there is flooding. It is best to create two disaster boxes, and give one to a physician and one to someone who is knowledgeable about the disaster plan, such as the office manager. Thus, if the doctor is out of town or incapacitated, your practice can recover without a significant delay.

The disaster box should be updated at least annually. Depending on how much your practice changes over time, you may decide to update the contents of the box more often. Bring all of the updated material to the location of the disaster box instead of bringing the disaster box to the practice. Even if you do decide to bring the box to the office, never leave it in your office building overnight. It is unlikely that the building will burn down, but it would be a great loss if it did and the disaster box was lost.

Because the disaster box contains very private and sensitive information, it should be kept secure. A plastic container with a watertight lid is a good option, as these are effective in keeping paper dry and intact during flooding. This kind of box is available from Wal-Mart or Office Depot. Consider placing a lock on your disaster box to ensure security and privacy of your data. Stronger disaster boxes,

Table 3-1 Risk assessment questionnaire for your practice.

1. Is your community or practice at risk for either a man-made or natural disaster?
2. Do you know what aspects of your practice need to be mobilized as quickly as possible so that you can see patients after the disaster has occurred?
3. Do your employees know what to do after a disaster so that they can care for their families and then return to work to help care for your patients?
4. Do you have a process that will allow you to communicate with your employees and physicians after a disaster if the usual method of communication is nonfunctional?
5. Are the bricks and mortar (i.e., the building where you practice) able to sustain a natural disaster? Have you made provisions to protect the equipment and patient records in your practice?
6. Have you made provisions to move your practice if you are unable to access your office or if your office building is unsafe for your employees or patients?
7. Will you be able to contact your patients and provide them with their records and important data even if you do not have access your office or cannot have access to your computers?
8. Does your community or region have a disaster plan to ensure the safety of its citizens?
9. Do you have adequate insurance, including business interruption insurance to sustain your practice?

If you answered *no* to any or all of these questions, then your practice is at risk if a disaster, either natural or man-made, impacting your practice or community and you will seriously want to put a disaster plan in place.

Disaster Box Checklist	Disaster Box Checklist
Updated on: _____ by: _____	Updated on: _____ by: _____
Organization Documents	Contact information
Articles of incorporation	Employees
Budget	List of patients
Most recent tax returns	Physician list
Administrative documents	Communication plan
Insurance policies	Inventories
Contracts	Computer / network inventory
Leases or deeds	Equipment inventory
Personnel policies	Office inventory
Employee manual	Photo or video inventory of your equipment and furnishings
Process documents (accounting procedures, hiring paperwork, etc)	Data backup
Financial statements / documents	Backup copies of software and licenses
Depreciation schedule	
Chart of accounts	
Bank account information	
Investment information	
General ledger	
Aged receivables	
Budget projections	

Figure 3–2 Disaster box check list.
Source: Disaster Preparedness, Continuity of Operations Planning for Nonprofits. Modified with the permission from the Louisiana Association of Nonprofit Associations.

such as an OrangeBoxx that contain holes to install a lock, are available from http://www.orangeboxx.com.

HUMAN RESOURCES

Regardless of the size of your practice, your employees should be protected from injury in your practice. (For a complete list of recommendations for your employees, see Chapter 10 on disaster planning for employees.) You must also consider how your employees and patients will be able to access your practice after the disaster has passed or cleared. The employee form (Figure 3-3) provides vital contact information so that you can maintain constant communication with each employee. With this information, you can have access to your employees and keep them apprised of the status of the practice, when they are expected to return to work, and where the practice is newly located. This is probably the most important form to be completed and maintained in your disaster recovery plan. You can have all of the physicians back in the practice and the building can be ready to see patients, but if you do not have employees, you are almost certainly paralyzed. After compiling all of your employee information, then assign a "call order" so that employees are contacted in the desired order, with the most important of your employees notified first, followed by those whose importance is less critical. Maintain a cur-

---

Name of employee: _____
Position: _____
Key responsibilities: _____
Home address: _____
City, State, Zip: _____
Home phone: _____
Cell phone: _____
Office phone: _____
Pager/beeper: _____
Home e-mail: _____
Work e-mail: _____
Emergency contact: _____
Relationship: _____

Figure 3-3 Employee contact list. Use this form to gather information on employees and your physicians so that each person can be contacted at any time or place. After you have entered all of your employees, assign "call order" for each employee. You may choose to sort your employee list alphabetically or by the order the employee is to be called. Maintain up-to-date contact information for each employee in an accessible and secure location. This example can be used for each employee as well as for each physician in the practice. This contact list should be updated on a regular basis, perhaps every 6 to 12 months.

rent list of phone numbers, and keep this list in an accessible and secure location; also, keep the list in multiple locations. Copies of this list must be distributed to physicians, office managers, and other key employees who will need to have access to the information.

Create a phone chain (or phone tree) that defines who calls whom. The responsibility of contacting every employee should not fall on one person. Contact information should include every known way of communicating with your staff and partners (e.g., phones, BlackBerry, pager, cell, and e-mail). An effective exercise is to test the phone tree on a Sunday evening to see whether you can get in touch with everyone.

Make one employee responsible for the telephone or e-mail tree. He or she would then be responsible for distributing the information. This important list should also be on the employee's home computer, as well as on his or her personal digital assistant. A hard copy should be kept in case there is widespread electric outage and an inability to use the computers at the office and home. The employee who is responsible for the phone list should update it regularly so that it is current and so that all current employees are on the list.

Another advantage of this employee form is that it allows you to know whom to contact if any of your employees become injured or sick at work. This information also is helpful if you need to contact the families of an injured or sick employee. Because it is necessary for employees to return to work before operations can be resumed, you might also want to consider the following:

- Alternate forms of transportation for employees (e.g., carpooling). Determine whether any employee has a four-wheel drive or van that could be used if the roads and highways are severely damaged.
- A contingency plan for emergency housing of displaced employees.
- Addressing immediate financial needs of your employees, including short-term financial aid.
- Childcare at your primary or alternate site (be sure to plan ahead with public officials to meet any regulatory requirements).

Payroll continuity is important for continued loyalty of your employees, as it helps them to handle disaster-related problems at home and to meet their personal financial obligations. You may want to establish a practice policy for the following:

- Direct deposit of paychecks for all employees
- Overtime pay during disaster recovery
- One week of pay, even if your practice is not operational

Plan ahead if you know that you will have to deal with security/access issues for your primary or alternate site. If employees need badges or security clearances, be prepared. Most disasters limit access to the disaster site, and emergency identification badges that all employees should carry in their wallets and purses may be helpful in achieving access to the practice or the hospital.[2]

The disaster plan should involve all staff members, as they are more likely to accept the plan if they have taken an active role in its preparation. Assign parts of the disaster plan preparation to each staff member, and encourage them to do the research on their assignment. Finally, a lagniappe (a Cajun word for "added benefit") also includes building a cohesive team and reassuring them that they will be cared for and that they can be assured of job continuity and security if a disaster occurs.

According to Debra Cascardo, an effective plan has a contingency if the initial plan cannot be applied.[3] Thus, if key members are unable to return to the practice, the practice can still be functional. In order for this to occur, comprehensive job descriptions must be created and placed into an electronic file, and a hard copy must be available to all of the key members of the disaster recovery team so that other staff members can fill in if key staff members are not able to return to work. These extensively written job descriptions also have the benefit of cross-training your staff and are helpful in the event that one or several employees are absent from the practice.

KEY CONTACTS

After employees, your next most important audience is your patients. Having access to the practice management system and the patients' records is essential to the success and future of the practice. The form shown in Figure 3–4 assigns responsibility for the patients' records, the backup tapes, and the daily schedule list. It also assigns responsibility for calling patients to reschedule appointments and notifying them when you will be back in operation or where they should go for emergency care. You may want to keep a list of patients' medications, such as chemotherapeutic agents or medications that need

---

Name of business or service: _____
Account number: _____
Password: _____
Materials/service provided: _____
Street address: _____
City, State, Zip: _____
Company/service phone: _____
Primary contact: _____ Title: _____
Primary contact phone: _____ Contact cell phone: ____
Primary contact pager: _____ Contact fax: _____
Primary contact e-mail: _____ Contact website: _____
Alternate contact person: _____ Title: _____
Alternate contact phone: _____ Alternate's cell phone: ___
Alternate contact pager: _____ Alternate's fax: _____
Alternate contact e-mail: _____
Recovery notes: _____

Figure 3–4 Key contacts. Use this form to list the key contacts for administration/management of your practice. Key contacts consist of those you rely on for administration of your practice, such as your bank, your lawyer, your insurance agent, and accountant. They also include services in the community you need to help you resume operations, such as utilities, emergency responders, media outlets, and business organizations.

Examples of key contacts include the following: accountant, gas/heating company, police department, bank, hazardous materials, public works department, benefits administration, hospital, small business, billing service, insurance agent, administrative office, building manager, insurance company (claims reporting), telephone company, building security, local newspaper, electric company, local radio station, local television station, emergency management association, fire department, and payroll processing.

to be refilled immediately, and when they should be given. With an electronic medical record program, you will find that most of these computer programs can provide this information. If not, check with your vendor to find out how this can be done. This vital information should be kept by at least one physician, the practice administrator, and should also be kept offsite in case you cannot obtain access to your building or your computers.

Next, you will need a list of your key contacts for emergency response and resumption of your critical practice functions. You will need to be able to communicate with the hospital(s) where you admit your patients and all of the insurance companies that hold most of your account receivables. Also, key contacts consist of those you rely on for administration of your practice, such as your bank, creditors, insurance agent, accountant, and lawyer. They also include services in the community you may need to access to help you resume operations, such as utilities (telephone, gas, and electric companies), emergency responders, emergency medical help, media outlets, and emergency numbers of the fire, police, and health departments. You should have well-established contacts with municipal authorities and other service providers before disaster strikes. The effectiveness of this liaison may be enhanced if a group or several physicians coordinate this communication for multiple practices.

PRACTICE OPERATIONS

The form in Figure 3-5 shows the list of the tasks that are to be done before the disaster and also includes the steps necessary to become operational afterward. This is a list of priorities that will need to be accomplished to minimize your losses and to open your practice after the disaster has passed. The form lists the functions that are to be performed by various staff members and assigns responsibilities for these functions with timelines for carrying them out. It also consists of several practice functions that must take place in order to be able to see patients again in the office. These functions should be

---

Practice function: _____

Priority: ☐ High ☐ Medium ☐ Low

Employee or physician in charge: _____

Timeframe or deadline: _____

Business function: _____

Priority: ☐ High ☐ Medium ☐ Low

Employee or physician in charge: _____

Timeframe or deadline: _____

Practice function: _____

Priority: ☐ High ☐ Medium ☐ Low

Employee or physician in charge: _____

Timeframe or deadline: _____

Brief description of procedures to complete function: _____
_____
_____

You should consider writing out two scenarios, one for a short disruption (i.e., several days) and the other for a more lengthy disruption (i.e., weeks or months).

Recovery note: _____
_____
_____

Figure 3–5 Critical business functions. This form identifies what practice functions are critical for survival. The following key questions are necessary for you to answer before the disaster:

- What are the most critical and time sensitive practice functions? Examples would be patient charts, practice management systems, and necessary equipment in your laboratory or radiology department.
- How much down time can the practice tolerate for each of these functions?
- What practice functions are necessary to sustain the financial obligations and maintain cash flow?

categorized into high, medium, and low priorities and then organized in a fashion so that the high-priority activities are fulfilled first.

Examples of these functions might include the following: Who will declare that a disaster has occurred and that the disaster plan should be implemented? Who will be responsible for the patients in the office at the time of the disaster and see that they are escorted to safety?

After the disaster supply kit has been created, the responsibility for it and the maintenance of its materials should be assigned to a staff member. The operations form should list the staff members who are essential and who are expected to stay in the practice as long as possible and as long as it is safe and to help with moving patients to safety, retrieving vital documents, and securing equipment and computers.

Whether your office has one computer or hundreds, after data are lost, they are almost always lost forever. There will never be easy resumption of your operations without data. One of the most important tasks to implement before the disaster occurs is properly backing up data; therefore, this form assigns responsibility for the vital task of securing the data and the task and the responsibility of verifying the data and regularly taking it offsite to a secure location. This can be as simple, like having someone take the backup home at the end of each day, or it could be high level, clustering or mirroring the server. The latter, however, is an expensive way to ensure data security but probably is more reliable than having someone in the office do it.

Who is responsible for managing the disaster plan if the disaster occurs after hours? It is not unusual for the local phone service to be disrupted—employees will need to be supplied with an out-of-area phone number to call for instructions about the future of the practice and what is expected of them.

Who is responsible for managing the deposits, and what should be done with the deposits if the bank is also not operational? Who is authorized to make bank transfers? Are there alternatives? Finally, who are the authorized check signers? This form should contain the names of the banks, their locations, contact numbers, e-mail addresses, account numbers, and who has authorization to enter the accounts and make transactions on the accounts.

For example, after Hurricane Katrina, one of us (NHB) had our deposits sent to another bank account in another city. Only after postal services were restored were the deposits placed in our bank in New Orleans. Because this contingency was not set up before the storm, it took several weeks to put this plan into action. Now, however, this plan of moving deposits to another location is in place, and there should not be any delay to have the deposits immediately sent to a second location, which is several hundred miles from the New Orleans area.

This form should also include the evacuation plan for the practice and a list of addresses for shelters. Also, the form should include accountability procedures and who is responsible to see that all of the tasks that have been assigned will be instituted.

The form should include the location of commonly used and necessary emergency equipment and who is responsible for removing the equipment. There should also be an assignment to check the emergency equipment regularly to verify that everything is performing as it should and that medications are not expired, resuscitation equipment is working, and that the fire extinguishers are in working order. Also, the plan should include the emergency shutdown procedures for computers and medical equipment and information about securing the facility by locking the doors and locking the cabinets to the medications and samples.

Activities that need to be instituted after the disaster include such administrative functions as follows:

- Recovery location setup
- Payroll and location of an alternative financial institution if your bank is also out of business
- Insurance claims (filing and following up on your claim)
- Regulatory requirements (e.g., time-sensitive reports)
- Debt obligations (bills due)
- Accounts receivable
- Communications—telephone lines, computers, and Internet access
- Restocking the medications and supplies necessary for your practice

COMPUTER EQUIPMENT AND SOFTWARE

The form in Figure 3-6 lists all of the computer hardware and software that are necessary to perform critical practice functions. Without access to data and information, your practice can come to

Name of Vital Record:		
Media:	☐ Network	☐ Print version
	☐ Hard drive	☐ Microfilm
	☐ Laptop	☐ Internet
	☐ CD	☐ Other
	☐ Diskette	Explain:
Is it backed up?	☐ Yes	☐ No
How often is it backed up?	☐ Hourly	☐ Quarterly
	☐ Daily	☐ Semi-annually
	☐ Weekly	☐ Yearly
	☐ Monthly	☐ Never
Where is it stored?		
Can the record be recreated?	☐ Yes	☐ No
Has the back up been tested?	☐ Yes	☐ No
Date of backup test? _____		
Practice function it supports: _____		
Recovery notes: _____		

Figure 3-6 Computer equipment and software form. It is for records that are vital to perform critical practice functions. Use "media" to indicate whether the record is a print version or is on a CD, diskette, and so forth.

a standstill. Possibly consider the services of a data center and disaster recovery facility, where data are backed up on a regular basis and available if your normal business operations are interrupted (see Chapter 6). Most practices are dependent on computers, using desktop and laptop computers and computer networks. Most communicate or conduct business with their hospitals, laboratories, vendors, patients, partners, and insurance companies through the Internet. The field of electronic communications is a rapidly growing segment of the economy. If your practice does the majority of its work online and uses computers, be sure that your computer security is current. This form needs the security codes to have access to the data if they are stored online and sent to an offsite server for backup. This form should also contain the source codes and vendor names and their contact information. This becomes important if your computers are damaged and you have to download your data to new computers during the recovery process.

If you move your practice to a recovery location, you will likely need to lease or purchase computer equipment and replace your software. Use the form to list what you would order. When there is sufficient warning about an event, such as a hurricane, you might decide to move some of your computer hardware and software to a safe place so that it could be used at your recovery location. This form provides you with that option. In addition, you could require all of your employees to take their laptops with them so that they can work from home or at the recovery location. Because some disasters occur without warning, you want to be sure that you have alternatives available (alternate site planning is discussed in greater detail in Chapter 9).

Other reminders for your computer and computer network include the following:

- Keep a backup copy of your computer's basic operating system, boot files, and critical software, and be sure that you have copies of your operations manuals.
- Maintain an up-to-date copy of computer and Internet log-on codes and passwords.

- Make arrangements with computer vendors to replace damaged vital hardware and software quickly and/or to meet your needs at your recovery location.
- Request written estimates for rental or purchase, shipping costs and delivery times, if relevant. Be sure to list these companies in your suppliers/vendors form as either primary or alternate vendors.
- Elevate computer equipment normally stored on the floor (e.g., CPUs), and secure in place whenever flooding is possible.
- Record the name and contact information for the company that provides website hosting.
- Record the name of the company that provides e-mail service, which may become highly important if you have to notify patients and staff of what is happening to your practice.
- Document all passwords that are needed to access files and data, and store these passwords offsite.
- Have an uninterrupted power supply, also known as a battery backup system, which will supply a limited amount of power in the event of an electrical outage. Ideally, server power switches and routers have power backups so that if a power loss occurs, you are able to shut down your network without causing damage to the server and other equipment. The website of American Power Conversion (www.apcc.com) has a resource to help determine what battery backup system is best suited to your equipment configuration. Monitors should not be plugged into the UPS devices because they will drain the power quickly from your battery backup system.
- Firewall drives are imperative for network systems that are always on. Without a firewall, you are opening your system to those who can hijack your site and have access to your data without permission.
- If you have a T1 line, if all phones, Internet, and e-mail services go through this line, and if the line goes down, your electronics will be rendered useless. If appropriate, make contingencies for this situation, such as setting up backup land lines for patients and/or staff use.

- Set up a free e-mail account (Hotmail, Yahoo, etc.) for emergency use. Document this, and share this e-mail address with your staff.
- Always keep your computer hardware and software licenses up to date.

This document should list every piece of hardware that your organization owns and would need to replace if damaged or destroyed. Include the make and model as well as the serial numbers of all of your hardware. Also document all printers and other peripherals (scanner, zip drives, etc.). Keep a book that has all of the purchase receipts with details of the hardware. Also, document all other technology equipment (i.e., phones, faxes, pagers, beepers, and cell phones).

Document all of the software being used in your practice. For Windows computers only, this documentation can be created by going online to www.belarc.com and running the Belarc Advisor, which builds a detailed profile of the installed software and hardware and displays the results in your web browser. Macintosh computer users can use the System Profiler, which is part of every Mac. Print these pages out, and store them offsite. This, however, has to be done for each individual computer.

Also have the contact information for the company that provides website hosting and that provides e-mail service. Of course, you must document all passwords that are needed to access files and data and store offsite.

VOICE/DATA COMMUNICATIONS

Use the form shown in Figure 3-7 to list your voice and data communications needs, including modem, voicemail, Private Branch Exchange (PBX)/Automatic Call Distribution (ACD), and standard telephone. Examples of data communications include cable, DSL, or dial-up for your Internet and e-mail access.

Communication with employees, vendors, customers, emergency officials, and other key contacts is paramount to your ability to

```
Type of Service:   ☐ Telephone              ☐ Fax machine
                   ☐ PBX w/ ACD*            ☐ Two-way radio & pager
                   ☐ PC data communications ☐ Other
                   ☐ Cell phone                Explain: _____
                                            _____

Description and Model Number: _____

Status:            ☐ Currently in use   ☐ Will lease/buy for recovery location

Voice Communication Features:
                   ☐ Voice mail             ☐ Conversation recorder
                   ☐ Speaker                ☐ Other
                   ☐ Conference                Explain: _____
                                            _____

Data Communications Features:
                   ☐ Cable                  ☐ Dial-up
                   ☐ DSL                    ☐ Other
                   ☐ T1                        Explain: _____
                                            _____

Quantity: _____
Primary supplier/vendor: _____
Alternate supplier/vendor: _____
Recovery install location: _____
Recovery notes: _____

*Automatic Call Distribution
```

Figure 3-7 Voice/data communications form. Use this form to list your voice and data communications needs. Communication with employees, vendors, suppliers, emergency officials, and other key contacts is vital to your ability to resume practice operations after a disaster event. This form should be used to determine what telecommunications equipment you need to help with that communication.

If you have to move to a recovery location, you will likely need to lease or purchase telecommunications equipment. You may use the voice/data communications form to list what you would order, and in the "Description and Model No." filed, write "unknown," or similar words, if you do not yet have that information.

resume business operations after a disasterous event. Voice and data communications equipment is only one requisite to a larger communications plan.

One of your critical needs is both internal and external communications; therefore, you can update your suppliers/vendors and key contacts, including your patients, on the status of your practice. This plan should include media relationships too. Although the demand for phone service may overwhelm the system, consider the following safety nets or alternatives as ways to communicate with your employees, patients, and vendors and/or suppliers:

- Designate one remote voicemail number on which you can record messages for employees.
- Arrange for programmable call forwarding for your main business line—if you cannot physically access your practice, you can call in and reprogram the phones to ring elsewhere. (After a disaster, telecommunications engineers are swamped with requests to redirect phones, faxes, and data lines to backup locations. If this is not in place, your recovery location could be affected.)
- Consider alternative forms of communication if phones are not working, especially to keep in touch with your employees. In anticipation of a break in all phone service, including cell phones, you might invest in some inexpensive two-way radios and pagers that send signals to each other over short distances. Several employees should know how to program phones to forward to another number, change voicemail messages, and retrieve voicemail and any other necessary phone communications.
- Communicate by e-mail, postings on your website, or on an emergency messaging system.

As you think about your voice communication needs at your recovery location, determine whether you need speakerphones, voicemail capacity, or the ability to record conversations. Also, decide whether you need a conference bridge to have conference calls with employees and key contacts to assess disaster damage and to make recovery decisions.

Here are some reminders:

- "Plain old telephone service," in which the handset is connected to the base, will likely work during a power failure. The connection is direct to the telephone company, which has extensive backup power.
- Cordless phones rely on electricity onsite and may be rendered useless in a disaster in which the power is out for extended periods of time.
- Cell phones may work if cell towers are still functional, but often system overload causes lost connections. This was the case with Hurricane Katrina.
- Consider small battery-operated cell phone chargers or power units for short periods of battery power to use when your battery runs down. These can be purchased at a home supply or hardware store and are available from Charge to Go (www.charge2go-sales.com) (Figure 3–8).
- Surge protection for all computer and phone equipment can prevent a power surge through a telephone line, which can destroy an entire computer. You may want to invest in a surge protector with a battery backup. You can find out what the surge protector is capable of by checking its Underwriters Laboratories (UL) ratings. UL is an independent, not-for-profit company that tests electric and electronic products for safety. If a protector does not have a UL listing, then it likely does not have any protective components at all. An inferior-quality surge protector may catch on fire and do more than damage your computer. This is actually a fairly common occurrence.

Many UL-listed products are also of inferior quality, of course, but you are at least guaranteed that they have some surge protection capabilities and meet a marginal safety standard. Be sure that the product is listed as a transient voltage surge suppressor. This means that it meets the criteria for UL 1449, UL's minimum performance standard for surge suppressors.

66 | **Chapter 3:** Preparing a Disaster Plan

Figure 3-8 Portable phone charger.

Surge protectors do not kick in immediately; there is a very slight delay as they respond to the power surge. A longer response time tells you that your computer (or other equipment) will be exposed to the surge for a greater amount of time. Look for a surge protector that responds in less than 1 nanosecond.

Better surge protectors may come with some sort of guarantee of their performance. If you are shopping for more expensive units, look for a protector that comes with a guarantee on your computer. If the unit fails to protect your computer from a power surge, the company will actually replace your computer. This is not total insurance, of course, as you will still lose all the data on your hard drive, but it is a good indication of the manufacturer's confidence in their product.

MISCELLANEOUS RESOURCES

The form in Figure 3-9 lists the basics to make your recovery site operational, such as office furniture, file cabinets, and temporary shelving.

ITEM	Quantity	Vendor/Supplier	Alternate Vendor/Supplier
Desks			
Chairs (reception)			
Cabinets			
Exam tables			
Chairs (exam room)			
Paper towel dispensers			
Wastebaskets			
Copy machine			
Fax machine			
Telephones			
Modem			
Surge protector			
Power strips			
Disposable gloves			

Figure 3-9 Miscellaneous resource form. This form lists all of the critical supplies that are needed to become operational and would be needed if you had to move your practice from your current site to another.

DISASTER RESPONSE CHECKLIST

When disaster strikes, you may be on your own for hours, several days, or even several weeks. Emergency services may not be able to respond right away. The checklist in Figure 3-10 includes supplies to help you take care of your employees, your patients, or others on your premises until help arrives. Your key contacts will include emergency services that you may need, such as the fire department, emergency management agency, and American Red Cross. You should be

- Water. If storage space allows, store 2 gallons of water per person per day for drinking and sanitation. Store in plastic containers or use commercially bottled water.
- Food and utensils. Have at least a 1- to 3-day supply of nonperishable food, which might include ready-to-eat meats, juices, and high-energy foods such as granola or power bars. Also include a can opener.
- NOAA weather alert battery-powered radio and extra batteries.
- Fire extinguisher.
- AM/FM radio (battery operated with extra batteries).
- Flashlight and extra batteries. Do not use candles or open flames during an emergency.
- Whistle to signal for help.
- Dust or filter masks. Readily available surgery masks will work fine and are available at your hospital.
- Moist towelettes for sanitation.
- Bleach. Use in the toilet if the toilet is not working.
- Basic tool kit, including wrench, hammer, and pliers to turn off utilities.
- Broom, shovel, and working gloves.
- Plastic sheeting and duct tape to "seal the room."
- Medications to include prescription and nonprescription medications such as pain relievers, antacids, and antihistamines.
- First-aid supplies, including an assortment of bandages, ointments, gauze pads, cold/hot packs, tweezers, scissors, hemostats, band-aids, gauze, nonadherent sterile pads (various sizes), paper and cloth tape, antibacterial ointment, burn cream, pocketknife (Swiss Army variety), razor blades, large cotton cloth (use for sling, tourniquet, bandage), nonaspirin pain reliever, chemical ice pack, hand warmer packets, safety pins (various sizes), needles, heavy thread, matches (waterproof), eye wash, hand wipes (antiseptic), cotton balls, cotton pads, alcohol swabs, and iodine (bottle or pads).
- Blankets.
- Battery-operated fans.
- Garbage bags and plastic ties for personal sanitation.
- Paper supplies, note pads, markers, pens, pencils, plates, napkins, and paper towels.
- Disposable camera to record damage.
- Cash/ATM and credit cards. Keep enough cash for immediate needs.
- Dehumidifier.
- Metal cart.
- Flashlights.
- 50-foot extension cord (grounded).
- Portable electric fan.
- Wet vacuum.
- Freezer or wax paper.
- Plastic trash bags.
- Plastic buckets and trash can.
- Paper towels.
- Sponges.
- Mop.
- Monofilament nylon (fishing) line.
- Broom.
- Gloves (rubber and leather).
- Rubber boots and aprons.
- Safety glasses.
- Plastic sheeting (stored with scissors and tape).
- Multi-KV generator.
- Safe or locked box.
- Copy of employee contact form.

Figure 3–10 Disaster response checklist form. Although this is not a complete list of all of the emergency supplies and equipment that you will need, it shows the most important supplies needed to care for the practice until help arrives.

able to put the basic disaster supplies together for under $100. One major purchase (more than $500) that you should consider is a multi-KV generator, prewired to the building's essential electrical current, which you can operate during a power outage for several hours.

DO YOUR EMPLOYEES KNOW ABOUT YOUR EMERGENCY PLANS?

Meet with your employees at least once a year to review emergency plans. Make sure that they know how to evacuate the building safely in an emergency and how to protect themselves and your patients in case of a disaster. Consider providing your staff with first-aid, CPR, and other emergency training.

Also, practice mock disaster drills. In addition to ensuring that employees know how to evacuate the building safely, make sure they know where to meet, to whom to report, when to leave the designated meeting place, and how or where to make contact if they fail to get to the meeting place. The law in most states requires that all high-rise buildings clearly post emergency evacuation routes out of the building in case of a disaster. Designate one employee to be the safety coordinator. Keep a list of emergency phone numbers—such as fire department, police department, ambulance service, emergency management agency—in a highly visible place (Figure 3-11).

VITAL RECORDS

Certain forms and documents will be required for you to perform critical functions for your practice. Figure 3-12 is a list of forms and documents that should be copied and kept in a secure, fire-proof, locked box. You should have a disaster box or kit that contains the most valuable documents, and this form determines responsibility for removing the box during an evacuation. Also, purchase a fire-proof, crush-proof safe box to store these crucial documents. Scan critical documents, and store them on a CD, on the Internet, or in password-protected section of your website. These documents can also be e-mailed to yourself and stored on your browser.

| Fire department _____ |
| Police department _____ |
| Building supervisor _____ |
| Local ambulance service _____ |
| Hospital (closest) _____ |
| Hospital security _____ |
| Hospital (alternate) _____ |
| Insurance provider/agent _____ |
| • Contact phone _____ |
| • Policy number _____ |
| • Headquarters phone/contact _____ |
| Telephone company _____ |
| Gas/heat company _____ |
| Electric company _____ |
| Water company _____ |
| Red Cross _____ |
| FEMA _____ |
| Radio station(s) _____ |
| _____ |
| _____ |
| Television station(s) _____ |
| _____ |
| _____ |
| Newspaper _____ |

Figure 3-11 Emergency phone numbers.

ARE YOUR EMPLOYEES PREPARED AT HOME?

Your employees are your most important asset. Your practice will not be able to function at your predisaster capacity unless your employees are safe and able to return to the practice. They will not be able to return to work unless their family needs have been met.

- Copy of 3 years of tax returns; 1 year of personal tax returns on principles (affiliates with greater than 20% interest).
- One year of tax returns on affiliated business entity (i.e., ancillary services such as computed tomography scanner, pathology labs, X-ray companies in which the practice is invested).
- For sole proprietorships: a copy of 3 years tax returns with Schedule C.
- List of creditors/contact information with account numbers.
- Sole proprietorships, corporations, and partnerships all need the following:
  - Copy of current profit and loss statement (current within 90 days)
  - Copy of listing of inventory
  - Copy of schedule of liability
  - Copy of balance sheet (as recent as possible)
  - Copy of all of your required licenses (city, occupational, sales tax, federal ID)
  - Copy of doctors' malpractice insurance
  - Copy of doctors' state licenses
  - Copy of doctors' medical school diplomas

Figure 3–12 Vital records. These lists show the vital records that you will need to have after a disaster for your practice to be functional.

Encourage employees to develop and exercise family disaster preparedness plans. Creating a disaster plan for your employees is discussed in greater detail in Chapter 10.

YOUR BUILDING

The building where you practice is important. If you own your building, you need to create a site map that indicates utility shutoffs, water hydrants, water main valves, water lines, gas main valves, gas lines, electrical cutoffs, electrical substations, storm drains, sewer lines, floor plans, alarm and sounders, fire extinguishers, fire suppression systems, exits, stairways, designated escape routes, restricted areas, hazardous materials (cleaning supplies and chemicals), and high-value items. Whether or not you own your building, you should create a list of emergency contacts such as your electrician, plumber, architect, and building managers. Documents regarding the building should be accessible to the appropriate personnel (office manager, building super, etc.) and available to them both onsite and offsite.

Examine your building for security weaknesses. On a regular basis, verify that the batteries for emergency lighting are checked regularly. Be sure that the stair treads have reflective glow-in-the-dark strips to aid in dark exits. Electric door/key pad locks should have a manual bypass cylinder lock. Also, check that the fire extinguishers are easily accessible and checked regularly. Educate the staff how to use the fire extinguishers, and test your emergency exit routes. Be sure to post emergency exit routes on the back of restroom doors.

If you own the building your practice occupies, have it inspected by structural engineers and contractors to determine its safety and the extent of the damage. If you do not own the building, work with the owner to have the building inspected. Whether or not you own the property, you are responsible for the safety of your employees, patients, and anyone else who may be on the premises.

PRACTICE MAKES PERFECT

After the plan has been constructed, practice is necessary. Just as fire departments and armies practice before they go live, your practice has to have practice and training before employees will feel comfortable with the plan and be able to execute it seamlessly. Your staff must know what to do; thus, a disaster preparedness and recovery plan should include employee training.

Practice of the plan includes a regular, perhaps monthly, data backup where you test for validity, functionality, and accuracy of the data restoration process. If you are going to do backups, you need to test them to ensure that you can actually restore the data. The disaster plan needs to determine how frequently you will test the backup system. Some practices actually bring down the entire system down and then restore it to see that everything is working properly. Disaster planning requires that you build emergency preparedness into the culture of the practice. Orientation sessions for new employees should include an overview of the contents and a copy of the preparedness manual.

Now that you have written your plan, your next task is keeping it current. Certainly the most difficult thing is getting started; the next is keeping it current. A plan that is several years old will probably not be effective as numbers have changed, employees have changed, and circumstances have changed making the older plan obsolete and not fully functional.

To verify that your plan is embraced by all your employees, including your new employees, build emergency preparedness into the culture of the practice. Orientation sessions for new employees should include an overview of the contents and a copy of the disaster plan in addition to the employee manual.

GETTING STARTED

These nine forms are available on the CD that is included with this book. You are encouraged to use it as you wish. You can cut and paste those sections that are applicable and expand those that need more elaboration. Assign staff members to complete the various sections. Take a copy home; store it on your home PC, and give copies to key personnel, including the physicians, office manager, and employees that will be responsible for implementing the plan.

Begin by assigning a team to help create the plan. Although small practices may be able to get by with just one person doing the work, larger practices will have to enlist the assistance of several employees, particularly in coordinating various portions of the disaster plan. For example, assign one team/person to complete the computer/technical portion and another team to complete the personnel portion. If appropriate, call this group the emergency management team to help provide some positive reinforcement and instill a sense of credibility for their efforts, particularly when this task is in addition to their usual responsibilities.

Next, assign some decision makers. Appoint a person or, if the practice is large, a disaster team that has the authority to make short-term emergency decisions (e.g., whether to evacuate the building). Then assign a chain of command. There has to be a chain and knowledge throughout the practice of who is in charge. In other words, who is second in command if the first person is not present or cannot be reached? These people should include those in leadership, but they should not be only the office managers or the physicians; however, if they are not the leaders, they must have the office manager or the physicians' approval. These people given the responsibility of being decision makers should be long-term employees or those who were part of the plan's preparation and who are familiar with the disaster plan. A guideline is that the person in charge remains responsible until the authority is delegated to another person within the practice.

Often, an issue for the people trying to create a plan is dealing with employee and even physician complacency. The bean counters may not want to spend money on tech-related systems that may never get used. One solution to this may be to outline the possible scenarios—what would happen if you do not have resources allocated and plans in place—and demonstrate the impact on the practice if a disaster occurs. We cannot emphasize enough that a disaster might force your practice to shut down temporarily, move its location to another office, or alter your day-to-day operations. If you are not prepared and do not have a disaster plan, the effect on your patients, your practice, and even your income can be devastating. We can think of no other aphorism that is more appropriate to this discussion than "an ounce of prevention is worth a pound of cure!" Translation: spend a few minutes, hours, and perhaps days to prepare for a disaster; then the pain of jump starting your practice will be much easier.

When creating a disaster plan, do not become overwhelmed by the large number of tasks ahead. Work on it in sections, doing first the things that seem most important—for example, personnel and computer/IT—and as time allows, add to your plan in incremental steps. The most important decision is to begin the process and to make some plans that can be implemented in the event of an interruption.

To write your plan, you must do some preparation, which allows you to get to the place where you then commit your plan to paper—you cannot write a plan until you do the preparation. The most difficult thing is getting started; the second is keeping the plan current.

Remember that you do not want to make the plan so dogmatic that it does not have flexibility and that it does not allow others to modify or to change it if the original plan does not work. The plan has to be able to be implemented without the person or the team that created it. It has to be legible, understandable, and able to be interpreted by even nonmedical personnel. If only a techie can implement your plan, it will most likely not be successful. Also, common sense must rule.

As things change within the practice—people come, people go, computer programs fold, new programs start—the plan has to be updated to reflect these changes. The ideal candidate for maintaining and updating the plan might be the person who oversaw the creation of the disaster plan or someone who was involved with the process.

BOTTOM LINE

No document or disaster plan exists that will address every situation and circumstance. Unfortunately, there are no cookie-cutter templates, and one size does not fit all. Some common elements exist among disaster plans, but each is unique and has different priorities and circumstances.

ACKNOWLEDGMENT

Chapter 3 is printed with permission by the Institute for Business & Home Safety as a derivative work of the *Open for Business*® toolkit at www.disastersafety.org.

REFERENCES

1. Centers for Medicare and Medicaid Services. "Health Insurance Portability and Accountability Act of 1996 Summary of Administrative Simplification Provisions." Available at: http://www.cms.hhs.gov/HIPAAGenInfo/Downloads/HIPAAlawsum.pdf, accessed February 11, 2008.
2. Institute for Business and Home Safety. *A Disaster Planning Toolkit for the Small to Mid-Sized Business Owner*. Available at: http://www.ibhs.org/docs/OpenForBusiness.pdf, accessed February 11, 2008.
3. Cascardo, D. C. "When Disaster Strikes: Getting Ready for the Next Big One," *Medical Practice Management*, 2006; 22(1):8–12.

Chapter 4

Implementing the Practice Resumption Plan

Perfect practice makes perfect.

—Cal Ripkin, Jr.

ACTIONS TO TAKE AFTER A DISASTER

Dr. Harris Evans, from the Internal Medicine of Long Beach, Mississippi, has clearly demonstrated the ability to resume practice operations, as he was one of the first physicians who treated patients after Hurricane Katrina. Having developed a disaster plan properly, the Internal Medicine of Long Beach was able to contact patients, begin follow-up on accounts receivable (because cash flow stopped for weeks), and treat patients rapidly during the aftermath.

KEY QUESTIONS

1. Are you prepared to resume your practice after a disruptive event?
2. Can you sustain a practice disruption for longer than 1 month?

3. Where will you get the resources that are necessary to resume practice?

4. Do you rehearse/update your disaster plan annually?

This chapter focuses on the execution of tasks that are necessary to resume practice operations, with particular attention toward patient, human, and financial resources. Preplanning activities in order to facilitate a smooth practice "reentry" are also addressed.

After a disaster has occurred, you will need to implement the plan. With proper planning, you can anticipate getting your practice started efficiently and seeing patients quickly after the disaster has passed. This chapter covers the implementation process and provides information about how to decrease the down time and the anxiety and uncertainty that is sure to accompany a disaster.

Included in this chapter is a list of issues to consider if your practice is severely damaged or destroyed. The list is not all encompassing but may provide an overall guide if a catastrophic disaster occurs. You may want to add, modify, or take out some of these items, depending on your practice and the extent of the disaster. You may also want to keep your checklist offsite, where it would not be damaged or easily accessed. The sample policies contained in this model disaster plan also provide steps regarding appropriate action in the event of a particular kind of disaster.

First, physicians and/or the office manager must make every effort to contact employees about the extent of the disaster, what action employees should take, and when the practice will reopen.

If your building is damaged, contact the landlord or the leasing agent, and then determine the extent of the damage. If your practice owns the spaces, contact your insurance agent or company as soon as possible to file a claim. If the damage is not significant, employees should be notified about when they are expected to return to work; however, the damage may be such that the practice will have to relocate permanently or for weeks or months. In this case, Chapter 9 discusses more about alternate or temporary space.

Next, give attention to your mail and phone service. Mail service should be rerouted immediately if necessary. Obtaining and rerouting your mail to a post office box may be the easiest solution: Either visit the local post office for a change of address form or access www.usps.gov. With a new address, you have immediate access to your insurance payments and other important written communication, particularly if your fax machine and Internet access were compromised. The practice should also consider where phone calls should be sent. Ask the phone company to place a recorded message on your phone, as this lets patients know the status of your practice and when you will be resuming operations. Also, tell patients where they are to go in case of an emergency medical problem. You may also want to let them know whether they can access to their medical records. Contact your phone service provider for options. A detailed voice message, such as this may be necessary: "The office of Dr. Smith is temporarily closed due to (explain disaster situation). You may contact us via e-mail at (e-mail address) or telephone at (temporary telephone number). We appreciate your understanding during this temporary office closure."

Next, make contact with your insurance agent or carrier to report a claim as soon as possible. When the insurance company is notified, give it a brief assessment of the damage. You should also take photographs or videos as proof of the damage. Inquire about how quickly you can have an adjuster sent to the location for a thorough evaluation of the damage to the facilities. The doctors, office manager, or person in charge of the disaster plan should request that the property be videotaped, thus ensuring that all of the damage is recorded.

Keep an accurate accounting of all damage-related costs. The practice and employees need to track all damage-related costs that are incurred. Examples of these costs might include mileage driven by employees, long-distance phone calls, equipment, mailing, leasing equipment, and so forth. Such costs are reported, with receipts, to the practice's bookkeeper or accountant, as these expenses are usually reimbursed by the insurance company. Appendix 12 contains forms for recording such reimbursable expenses.

We suggest that you keep all damaged goods on the premises until an insurance adjuster has viewed the property. After the premises are safe to enter, the practice then assigns personnel to conduct salvage operations as soon as possible. Any items or equipment that can be saved should be removed, but damaged goods should be kept onsite until seen by an insurance adjuster. If the equipment can be used again, it should be protected from further damage. For example, if the exam tables can be reused, photographs should be taken before moving them to the new/temporary location. This verifies their condition as well as the fact that they were physically present in the office.

To get your staff functioning as quickly as possible, conduct a meeting with your key employees, such as the office manager, the nurses, and the physicians. This meeting may be held at a location away from the practice if necessary. Although the following list is not all inclusive, it suggests some topics that might be discussed at the meeting:

- Damage assessment
- Status of employees
- Medical records access
- Financial resources
- Information processing
- Office space needs—temporary/permanent
- Immediate equipment needs
- Contact of patients and suppliers

If there is significant damage of the equipment in the office, you will need to consider equipment needs for temporary office space. What will be needed, both in the short and long term, if office equipment is damaged or destroyed? To have an estimate of what might be needed, the practice should consult the already recorded list of equipment and assets (this was described in Chapter 3 and Appendix 6). This list is also important for insurance purposes.

After the extent of the damage and priorities are determined, the practice's patients should be contacted. Depending on the circumstances, patients should be told about the damage and where to find information about treatment and patient records. This can be done by advertising in the newspaper, on the radio, or with a mailing (this mailing list should have been kept offsite). Furthermore, the practice might use various websites, such as those operated by the media (television, radio, and newspaper) and local, regional, and state Medical Society agencies. Also, the practice website (if available) is invaluable during these situations, as it apprises patients about the status of the practice and provides information about how to contact the physician for prescription refills, medical records, and so forth. Consider contacting other medical providers to find suitable alternative care for your patients. Perhaps other physicians will be able to see your patients until your practice and new facility are operational?

IMPLEMENTATION PLAN

Following the guidelines and steps in this chapter will help managers to create a formal practice-resumption plan. It will help your practice become better prepared to execute essential business processes if an incident occurs.

As you address the issues that follow, ask yourself these questions: What do I do when we cannot use our facility? What can I do now to better prepare my practice to respond if our facility is unavailable? The reason—a fire, a tornado, or a massive power outage—for the unavailability is not the issue. Consider that your offices and all of the resources you have for day-to-day operations are no longer accessible.

PREVENTION

Identify potential problems, and plan corrective action. Address those issues that you can solve and those that will benefit. Some areas to recognize include the following:[1]

- Maintain good general housekeeping. Keep areas clean and free of obstructions and fire hazards. Remove any stored paper from common areas and store in restricted areas. Consider implementing a "clean-desk policy." In the same way that a large-city phone directory does not burn as easily as loose paper, removing loose paper from desktops to files at

the end of the work day can reduce losses from fire. This will also help to protect those documents from sprinkler discharge and other incidents.
- Look for and eliminate any obviously overloaded electrical circuits. Employees may have installed nonbusiness electrical appliances such as coffeepots, radios, space heaters, and fans. These appliances can cause electrical fires by shorting out or overloading circuits not designed for these appliances. Your building maintenance staff may be able to help you educate your staff about the problems related to these appliances.
- Observe physical security procedures in your facility, and encourage increased security when appropriate. Ask yourself these questions: Is your building open to the public? If you have restricted access, how and when can the building be accessed?
- Observe information security procedures regarding computers in your facility, and encourage increased security when appropriate. Ask yourself these questions: Do your staff members have their passwords taped to their monitors? Are your laptop computers secured at the end of the workday? Do your staff members leave their computers logged onto the network when they are away from their desks?

You may not have direct control over some of these prevention steps, but you can and should encourage those who do have authority to take the appropriate action. Consider scheduling security-training sessions on a regular, perhaps semiannual, basis.

PLAN DEVELOPMENT CHECKLISTS

The checklists contained in the appendices are invaluable during the practice resumption implementation phase. The checklists used at this time include the following:

- Corporate headquarters telephone numbers (Appendix 10)
- Practice recovery work area checklist (Appendix 11)

- Resources required over time (Appendix 12)
- Travel request form (Appendix 13)
- Recovery boxes (Appendix 14)
- Critical resources to be retrieved (Appendix 15)
- Personnel location control form (Appendix 16)
- Status report form (Appendix 17)
- Activity schedule (Appendix 18)
- Guide to record retention (Appendix 19)

These checklists, as found in Appendices 10 through 19, should be prioritized. For example, the personnel location control checklist should be initiated first in order to coordinate all activities among practice employees and to establish each individual's responsibilities. Additionally, certain employees should be responsible for patient notification, vendor notification, and so forth.

PATIENT LIST

After prioritizing the practice resumption plan, first contact patients with scheduled appointments and those who need and expect personal notification. Include those patients who would take their business elsewhere if not contacted. You must be proactive in contacting patients in order to mitigate any ill will or future business losses.

Next, various business partners (referring physicians, vendors, etc.) should be contacted so that regular operations can be reestablished. Be sure to notify the referring physicians about the resumption of practice activities, as they will have patients who need specialty services. Furthermore, vendors are crucial during the practice resumption phase, as they help to obtain equipment and supplies and coordinate other necessary resources. Some vendors may not have 24-hour service; if, for example, your incident occurred on a Sunday afternoon, contact the vendor directly via cell phone or e-mail, and discuss your concerns with the representative. Because many disruptive situations occur after normal business hours, 24/7 contact information must be available for all critical resources and must be part of the practice disaster plan.

PRACTICE RECOVERY STEPS

During a real disaster, some or all of the following steps might be implemented. The team leader will need to use his or her judgment while managing the practice resumption operation. The team leader must contact the emergency management team to find out when voice communications will be available, when service will be operational, how current the master files are, and whether other communication will be available (i.e., telephones, e-mail, and Blackberry).

Next, a meeting with key practice personnel will help to determine appropriate actions and to establish the priorities for restoring practice functions based on the work area and resources available. The practice manager needs to explain the goals and the objectives identified by the emergency management team as follows:

1. Review tasks to be performed, and assign personnel.
2. Personnel should notify patient/clients and vendors about the situation and about when service is to be restored.
3. Determine whether some personnel will have to travel to the practice recovery site.
4. Distribute copies of any forms that will be needed during the recovery operation.
5. Distribute copies of the newly prepared news media statement. Copies can be obtained from the emergency management team. Instruct team members not to make statements to the news media.

Determine which personnel are needed to resume normal practice functions at the work area. After contacting personnel who need to report to the assigned work area, designate space for all personnel, and establish procedures to resume time-dependent functions based on the established priorities. Furthermore, instruct all practice personnel to carry photo identification with them at all times and to be prepared to show it to security or local authorities. As progress continues during the recover operation, the team should be prepared to move back to the affected facility and resume normal practice operations.

The practice manager or assigned individual, after activation of the practice resumption plan, will need to contact all practice personnel using these procedures:

1. Place a telephone call and ask for the individual. If available, provide the necessary information.
   - Remind the person to make no public statements about the situation.
   - Remind the person not to call coworkers (unless instructed to) and to advise their family not to call other employees.
   - Record the information in the contact status column.
2. If not available, ask where the individual can be reached.
   - If the person is somewhere besides the practice center, get the phone number. Then call and provide the necessary information.
   - If the employee was working at the practice, indicate that you will attempt to contact the employee there. *Do not* discuss the disaster with the person who answers the phone.
   - Immediately notify the emergency operations center.
   - Record the information in the contact status column.
3. If contact is made with an answering machine, make no statement about the situation. Provide the phone number for emergency operations center, and ask for the employee to return the call as soon as possible.
   - Record the information in the contact status column.
4. If there is no answer, record that a call his been made in the contact status column.
5. If there is no answer and the individual has a beeper, place a call to the beeper number. Then enter the number of the emergency operations center, and record the information in the contact status column.

After contacting the appropriate personnel, give the following information to each team member:

- A brief description of the problem
- The location of the emergency operations center and/or the business recovery site

- The phone number of the emergency operations center
- Immediate actions that are to be taken
- Whether the facility can be entered
- The location and time that the team should meet or scheduled conference call
- The fact that all team members should always carry photo identification and be prepared to show it to security or local authorities
- The fact that no statements are to be made to the media

All callers need to record the status of the calls, noting the time the call was placed and whether contact was successful. Make a reasonable number of attempts, two or three, if the phone was busy or if there was no answer. Forward the completed list to the practice manager, as the staff should continue to attempt to contact team members.

GUIDELINES FOR TRAVEL TO THE PRACTICE RECOVERY SITE

Most disasters are isolated to a single building or block. During those situations, the practice recovery site in the local area will be used for recovery. Some disasters are community wide and, as such, may eliminate the option of using the local practice recovery site. In those instances, you may resort to using more distant recovery sites.

The practice manager should divide the available personnel into two groups: those who will go to the backup site first and those who will be sent as replacements after a few days. The department leader should not overcommit resources during the first few days.

The practice manager will have to provide directions to personnel who will be traveling to the backup site. If personnel cannot drive to the backup site and will need air transportation, hotel accommodations, and advance expense money, the team leader should arrange the details through the physician or practice manager. The practice manager will make the travel arrangements and will provide personnel with itineraries, tickets, and advance travel money when necessary.

OFFSITE STORED MATERIALS

At this time, the practice should access the "recovery box," which contains specific items for reestablishing practice operations. These items might include the following:

- Copies of necessary forms
- Copies of medical records
- Copies of business records
- Copies of procedure manuals
- Inventory listing (with pictures, invoices, etc.)
- Insurance policies
- A small supply of unique necessary supplies
- Backup tapes

Because the recovery box is in an offsite location, a designated individual needs to have access to the recovery. Furthermore, other critical documents, equipment, and supplies will be needed at this time, either from offsite storage or selected vendors. The recover box and all of its contents must be stored in commercial storage facilities in a climate-controlled and secure area. The practice should also consider a second "mobile" recovery box, which the physician and/or practice manager would have.

CRITICAL RESOURCES TO BE RETRIEVED

Many disasters do not completely destroy contents of offices. Depending on the circumstances, even if computer diskettes, tapes, and hard drives have been water, smoke, or soot damaged, it might be possible to extract the information from them. Do not attempt to do this yourself. Instead, contact your technical support or facilities staff for help.

After the incident, you might be allowed, depending on safety, into your building. This could be for as little as 15 to 30 minutes. Create a list of critical items that need to be retrieved. This assumes, of course, that the items are salvageable. Examples might include computer disks, computers, selected paper files, and work in progress. Examples of items that are not essential include personal items, unimportant files, and duplicate information.

The key to the practice resumption plan is preparedness. The practice disaster/recovery plan should be reviewed and updated on a semiannual basis and should include the following areas:

- Team leader and alternate team leader
- Recovery team alert list
- Critical functions list
- Team recovery steps
- Patient list
- Vendor list
- Work area requirements
- Offsite stored materials

TRAINING/EXERCISE SCHEDULE

One of the key components of the practice resumption plan involves selecting an alternate practice meeting location. Make sure that key people know the location and have maps if necessary. This predefined meeting place will be where all personnel plan their response to the incident. In choosing this meeting place, think about any necessary key resources, and consider the reliability of the physical location. Some of the resources and location considerations are as follows:

- Location. When selecting your meeting place, consider its location relative to your normal work place and to the key staff members. The location should not be so far away that staff members have difficulty getting there. Conversely, it should not be so close to your normal work location that it could be affected by the same incident. For example, after certain incidents, authorities may block several city blocks around the affected facility. If your meeting place is across the street from your normal work location, access may be unavailable.

- Alternate meeting place. To solve this issue, select at least two possible meeting locations. Your primary location could be close to your facility and could be used if access is possible. Your alternate location should be further away, ensuring availability if your primary location is not accessible.
- Vulnerabilities. When selecting a location for your meeting place, especially for your alternate location, be sure to consider the types of vulnerabilities. For example, your meeting place should be inland. If your primary location is near a river, your meeting location should be on high ground. If, on the other hand, your primary location is near an earthquake fault, your meeting location should a reasonable distance away from the fault line.
- Communications capability. Because the ability to communicate with others is essential for an effective response, make sure that the chosen location has enough telephones. If you have a cellular phone, take it with you as another means of communication, as regular phones may not be working. If you have a portable/laptop computer with Internet or e-mail capabilities, the meeting place should be able to connect that computer as well. If your laptop computer is available, plan to take it to the meeting place.
- Size of the facility. The location should be big enough to congregate. This is not an alternate place for your staff to work, however—it is only a place to discuss your plan of action and to manage your recovery efforts; therefore, it does not need to be big enough for your entire staff to work. The alternate work location will come later when your complete plan is documented.

When selecting a meeting place, the most viable alternatives are local hospitals or a hotel or convention center, as these facilities typically have emergency power resources. Advanced contact should be made to verify that the practice can use these facilities during an emergency situation. Furthermore, obtain their consent in writing and include the name, street address, telephone numbers, and security requirements necessary to activate access. Also, consider adding a map of the location and a floor plan of the facility for the office staff's use.

RESUMPTION ACTIVITIES FOR INFORMATION SYSTEM CAPABILITIES

The heart of your medical practice is often tied to your computer system. Computers can fail for many reasons. Disasters such as fire, floods, hurricanes, and computer hardware failure can place your practice in a vulnerable state. Copying data and backing up your system is a valuable and cost-efficient part of your practice. Establish and implement procedures to back up and maintain your patient/financial data, as this will help your practice to resume normally if a disaster occurs.

There are many ways to protect and restore your computer system data, depending on the size of your practice. Small practices might back up their systems on CDs, disk, or tape drives and store them at a remote location. External hard drives enable you to confidently and reliably store, manage, and protect your critical business information in a one-step backup. Powerful, yet easy to use, external hard drive products are designed specifically for small- and medium-sized businesses. After your backup is complete, take the external hard drive offsite with a complete data backup. In addition to system backups, your practice should remove all purchased software (i.e., QuickBooks and Microsoft Office) to reinstall if computers are left onsite and become damaged. Larger practices may choose to purchase expensive high-tech systems to control the backup process. Regardless of your practice's process, you should always consider the following:

- Are we backing up our data?
- How often should we back up data?

- What should we back up?
- What restorative process will we use?
- Where do we store our back up data?

Reports that should be stored offsite include the following:

- Accounts payable (month-end report)
- Accounts receivable (various report formats, such as a detailed aged trial balance by payor/by patient) total, insurance companies, and by patient
- Patient/client listing (with contact information)
- Insurance policy information and contacts
- Vendor contact information

In the event of a large-scale disaster, compile a listing of key external resources as found in Appendix 15.

BOTTOM LINE

The importance of planning for practice resumption activities and alternatives cannot be overemphasized. Everything from telephone and mail service to temporary office space to possible replacement of employees must be considered. Furthermore, particular emphasis should be given to the methods by which key contacts can be communicated with and how the practice can marshal the critical resources necessary to resume practice.

REFERENCE

1. Arlington Economic Development. Available at: http://www.arlingtonvirginiausa.com/docs/BizResumptionTemplate.doc, accessed February 11, 2008.

Chapter 5

Protecting and Recovering Practice Assets

An ounce of prevention is worth a pound of cure.

—Benjamin Franklin

Dr. Brian Anthony, a general surgeon with the Beach Surgical Group in Bay St. Louis, MS, purchased a trailer shortly after losing his medical practice from Hurricane Ivan. He did not realize the value of this asset until the threat of Hurricane Katrina, when he was able to salvage most of his patients' charts, computer equipment, computer hard drive, important papers and documents, ancillary medical equipment, and critical medical supplies. Because of his disaster planning and the ability to implement his plan, Dr. Anthony was able to reopen his practice in a temporary building and begin seeing patients, as he had his resources available.

KEY QUESTIONS

1. Do you have backup of critical patient and financial data, and is it stored offsite?
2. Do you have a records-retention policy?

3. Do you store key practice assets in a safe location?
4. What is your plan for temporary office operations if your office is unusable?

In this chapter, we discuss the development of information that is necessary for resuming practice activities. We learn about critical resources and assets that should be identified in case of a disruptive event.

The protection and recovery of key medical practice assets ensure that a practice can resume operations as quickly and seamlessly as possible after a disruptive event. Most key assets within a medical practice have vital data information (i.e., patient/financial/key contact information) that is necessary for continuing to practice immediately.

This chapter focuses on developing various lists of critical information, especially electronic resources, to help to reestablish practice operations. These data can include paper records, database records, e-mails with attachments, voice mail, instant messages, or any other official record documentation. Although proper protection starts with proper planning, the recommended approach for protecting the practice's vital records against all disasters (natural or man-made) are as follows:

- Designate a vital records program custodian. One individual, typically the practice manager, should be held accountable for planning and maintaining the vital records program. His or her job description should delineate the responsibilities regarding records management, business continuity, risk management, and emergency preparedness.
- Assess your current vital records program. Determine which records already exist and how these records can be assimilated in the event of practice disruption.
  - Medical records management (i.e., patient information)
  - Business continuity (i.e., accounts receivable/accounts payable records)
  - Risk management (i.e., insurance policies)
  - Emergency preparedness (i.e., information necessary to reestablish business functions)

In order to identify and access vital records, you must conduct a review of business operations and document the appropriate records that relate to the functional areas of the practice. All key information that is on paper should be scanned for safe keeping into the "key documents" section of the information system. Also, additional hard copies should be kept with the practice's disaster recover kit.

An initial plan to protect and recover vital records should include those records that are necessary to implement the business continuity and recovery plan. They should be easily accessible. Also, these records should also be periodically reviewed and updated to ensure effectiveness, and the vital records program should become a part of the overall compliance plan of the practice. This ensures the periodic review and updating of this critical information. Clearly, the management information system of any medical practice contains huge amounts of irreplaceable data and, effectively, becomes the most critical part of any disaster recovery program.

Information about all hardware and software associated with the practice, including serial numbers, original cost, replacement cost, access for backup data, as well as a plan for offsite data storage, should be documented. Furthermore, consider using Internet-based management information systems/electronic medical records in disaster-prone areas. Also, determine where, if applicable, the offsite server is located. For example, during Hurricane Wilma, servers located in south Florida became inaccessible to clients across the United States. Those practices that did not have their key documents recorded and placed in a secure location were at a distinct disadvantage compared with those practices that had their documents backed up and safely maintained offsite.

Concerning the protection and recovery of electronic data, it is important to consider the following items:

- Complete a full backup of all critical files on a daily basis.
- Backups can be set to run automatically, usually during the evening/nighttime hours, and daily backup ensures that all work data are up to date through the end of the previous workday.

- Assign responsibility for mandatory daily data backup.
- Review the backup software logs on a periodic basis in order to ensure compliance.
- Regularly test data backup logs to ensure the completeness of these processes.
- Identify offsite storage locations.
- Do not forget data on desktops, laptops, and PDAs.
- Ensure that open files are being backed up on a daily basis.
- Create written instructions for restoring backup data.
- Identify other hardware/software-related issues.
    - Identify sources of rental/loaner computer equipment and service personnel.
    - Use uninterrupted power supplies (UPSs and surge protectors, especially on servers and phone systems).
    - Use antivirus software, and keep virus definition files up to date.
    - Implement appropriate security on all servers and networks.
    - Use firewalls at any external network access point (i.e., Firewalls built into an external source, as in a router, also known as "hardwired firewalls," which provide a good front line of defense to the fact they are not a software firewall. External firewalls [hardwired] are encoded into a chip to provide superior protection at a minimum cost).

    A hardwired firewall is a physically independent network device that connects between your office network and the outside world (i.e., computers outside your office network). It acts like a traffic cop at a VIP event, allowing only safe, legitimate traffic through, while blocking access to unwanted traffic.

    A "soft-wired" firewall is a software application that resides on the local computer(s). Although it protects the local system, it does not prevent attacks on your Internet link, and it is thought of by most experts as a good supplemental layer of protection, but not a comprehensive solution.

Protecting and Recovering Practice Assets | 97

- Back up all financial and billing information daily.
- Develop secure list of employee passwords, including computer log-in, e-mail, voice mail, information system access, electronic medical records access and other software applications, such as QuickBooks, or other local or national payroll companies.
• Retain blank business checks in a secure offsite location.
• Ensure that contact information and passwords are available through the Internet service provider in order to access and/or change information on the practice's website as well as be able to access e-mails from various locations.

The key information that should be documented, should be stored in various locations, and should be easily accessible in order to resume practice operations includes the following:

1. Accounts payable information
2. Accounts receivable aged trial balance by payor/by patient
   a. Ensure that detailed information is updated at least weekly and is stored offsite.
   b. This information should contain patient name/address/phone number/key contact information for the insurance company.
3. Banking information
   a. Have account numbers/routing numbers.
   b. Consider the use of regional/national banks versus local banks.
   c. Establish electronic banking arrangements.
   d. Ensure electronic deposits from all third-party payors who offer such services.
   e. Make arrangements for employee direct deposit.
4. Building information
   a. Have external/internal security necessary before and after a disaster.
5. Electronic medical record access information
6. Employee listing

7. Insurance policies
8. Internet service provider
9. Patient listing
10. Referring physicians listings
11. Vendor information
12. Key contact information regarding the following:
    a. Accountant
    b. Attorney
    c. Cable/Internet service provider
    d. Electric company
    e. Fire department
    f. Gas company
    g. Health department
    h. Heating/ventilation/air conditioning service provider
    i. Hospitals
    j. Insurance policies/agents
    k. Landlord/property manager
    l. Media
       - Television
       - Radio
       - Newspaper
    m. Local/regional/medical societies
    n. Office service providers (i.e., housekeeping, biohazards Waste, etc.)
    o. Payroll company
    p. Pharmaceutical representatives
    q. Plumber
    r. Police
    s. Post office
    t. Practice management consultant
    u. Security
    v. Telephone company
    w. Water treatment resources

After assimilating this information, critical documents storage and accessibility are imperative in resuming practice operations. These documents should be stored in an offsite location(s) (preferably more than one) by key practice personnel (physicians and/or practice manager). They can be stored on CD, Internet, and servers and may even be held in the password-protection section of the practice website.

Other considerations regarding the protection and recovery of assets include the following areas:

1. Use temporary office space.
    a. Temporary office building (i.e., GE Medical Buildings)
    b. Temporary space at hospitals
    c. Secure locations within a 30-minute radius of the practice
       - Temporary office space
       - Sharing office space with colleagues
2. Explore records storage companies.
3. Consider offsite equipment storage companies.
4. Develop equipment specifications inventory/listing.
    a. Manufacture
    b. Model
    c. Serial number
5. Consider vital instruments/supplies/pharmaceuticals to be stored offsite.
    a. Obtain plastic storage boxes for quick mobility.
6. Consider mobile covered trailer (if time allows, such as a hurricane).
    a. Check with your CPA to see whether the purchase of a mobile covered trailer would be an allowable practice expense.

Finally, the practice should set up a records-retention policy for all vital records (e.g., patient information, financial records, and tax returns).

The guide to record retention should read as follows:

Medical Records*

Patient Charts . . . . .Permanently
Patient Charts—Alternative (adults)   . . . .Ten years after the most recent encounter

continues

Patient Charts—Alternative (minors) . . . . Age of majority plus statute of limitations
Medical Correspondence (to patients, referrers, etc.) . . . Permanently with chart
X-rays . . . . Permanently with chart

Other Medical Record Issues

Patient Requests Transfer

When transferring medical records, the physician should maintain the original record and should transfer only a copy. You may charge the patient a reasonable fee to reflect the cost of the materials used, the time required to prepare the material, and the direct cost of sending the material to the requesting physician (note: this may be determined by state law).

Physician Relocates

Physicians relocating their practice may take the medical records with them or leave the records with a designated custodian with an agreement that they will be permitted ready access to them as required in the future on request.

Physician Ceases Practice

If a physician ceases to practice medicine, he or she may be obligated to either transfer their patients' records to another physician at a local address and phone number or notify each patient that their medical records will be destroyed in (state specific) ___ years, unless they collect the records or request a transfer to another physician within 2 years. You may wish to contact your liability insurer for additional guidance.

Medical Records in a Group Practice That Is Changing

Physicians in a group practice setting usually have an arrangement that clarifies ownership of the records and have a records-transferring policy. Despite these arrangements, any physicians in any setting (e.g., solo practice, group practice, or hospital) are ultimately individually responsible for their own patient records. Physicians must be aware that agreements made with their associates do not supersede their responsibility to patients.

Typically, most physicians in a group practice arrangement will have an agreement with their associates that addresses such items as

- The method for dividing medical records on termination of the practice agreement. This usually specifies how to determine custody of the medical records.
- Some reassurance that each physician will have reasonable access to the contents of the medical records for preparing medicolegal reports, defending actions, or responding to a complaint investigation.
- Often, if no such agreement exists, physicians dissolving their joint practice try to agree on a system that determines "the most responsible physician" for each record. For example, the physician who has created the greatest percentage of the entries in a particular patient record may be expected to continue maintaining it.

- Although this approach is customary in most group practices, it is not mindful of the patient's needs. See details in the following section.

Ask the Patient

Members of a group practice must be cognizant of the fact that it is the patient's right to choose which doctor maintains their particular patient records and provides continuing medical care, regardless of the existence of an agreement.

A copy (or original) of that patient's records should be transferred, and physicians should agree how the cost of copying and transferring records will be divided. In the case of planned group practice dissolution, the cost cannot be charged to the patient.

Unexpected Dissolution of a Group Practice

Unexpected dissolutions of group practices create special difficulties. Ideally, the physicians should amicably agree on a strategy for informing patients and dealing with the medical records. In the case of a sudden, unforeseen departure of a partner or associate, records should be kept at their current location until the patient directs where he or she wishes to receive his or her ongoing health care. Reasonable access to medical records must be given to all former partners and associates.

Statutory Requirements

There are some statutory requirements on the keeping of medical records. For example, certain Medicaid/Medicare reimbursement regulations require that the medical records of recipients be available for review for 7 years.

Tax and Financial Records*

Record	Retention
Accounts Payable Ledger	Permanently
Accounts Receivables Ledger—Annual	Six years after the due date of the practice tax return
Accounts Receivable Ledger—Monthly	Two years
Bank Statements with Canceled Checks	Six years after the due date of the practice tax return
Capital Asset Records	Six years after the due date of the practice tax return for the year in which the asset is disposed
Cash Recipients Journals	Six years after the due date of the practice tax return
Check Register	Six years after the due date of the practice tax return
Daysheets	Six years after the due date of the practice tax return
Deeds, Mortgages, and Bills of Sale	Permanently
Deposit Books and Slips	Six years after the due date of the practice tax return

continues

Depreciation Schedules .... Permanently
Encounter Forms .... Six years after the due date
 of the practice tax return
Financial Statements—Annual (year end) .... Permanently
Financial Statements—Periodic .... Two years
General Ledger .... Permanently
Income Tax Returns (correspondence and audits) .... Permanently
Income Tax Returns (federal and state) .... Permanently
Insurance Policies (expired) .... Three years
Insurance Policies, Current Accident Reports, Claims, Policies, etc. ...Permanently
IRA and Keogh Plan Contributions, Rollovers,
 Transfers, and Distributions .... Permanently
Paid Invoice Expenses .... Six years after the due date
 of the practice tax return
Payroll Ledger .... Six years after the due date
 of the practice tax return
Payroll Tax Returns .... Permanently
Petty Cash Vouchers .... Three years
Stock and Bond Certificates (canceled) .... Seven years
Vouchers for Payments to Vendors, Employees, etc.
 (includes allowances and reimbursement of employees,
 officers, etc., for travel and entertainment expenses) .... Seven years

*Many of these documents are obviously maintained electronically. We recommend downloading this file to a disk or CD for storage, as indicated.

Employer

Employee Personnel Records (after termination) .... Seven years
Employment Applications .... Three years
Time Cards and Daily Attendance Reports .... Seven years

Other

Accident Reports/Claims (settled cases) .... Seven years
Correspondence, General .... Two years
Correspondence, Legal and Important Matters .... Permanently
Correspondence, Routine with Customers or Vendors .... Two years
Minute Books of Directors, Stockholders, Bylaws, and Charter .... Permanently
Trademark Registrations, Patents, and Copyrights .... Permanently

*State guidelines vary. Check with a local medical records training program, your professional liability carrier, or your health care attorney.

Source: Reprinted with permission from Gates, Moore, and Company.

Indeed, although the planning process to ensure the protection and recovery of key practice assets seems voluminous, this effort can be appreciated only in the case of a disruptive event, as preparation will pay off both in terms of being able to offer patient care services in a prompt manner as well as ensuring the successful preparation and financial recovery of the practice as it returns to normal business operations.

BOTTOM LINE

In order for a medical practice to resume operations, the practice must develop key contact lists, assimilate vital records (patient, financial, etc.), and have immediate access to these resources. Also, the practice should determine its information system capabilities and alternatives. In the next chapter, we discuss the backup and retrieval of electronic data, as this is vital to the operation of any medical practice.

Chapter 6

Creating a Backup Plan

To risk losing your data means you are at risk of going out of business.

—Sam Gutman, President and CEO of Intronis[1]

Dr. Floyd Buras, a pediatrician in eastern New Orleans, lost his practice in the flooding caused by Hurricane Katrina. His office was consumed by 8 to 10 feet of water, and his medical records were like paste when he returned to his office a few days after the storm. As a result, he had tremendous difficulty in helping his patients; however, with the assistance of his staff and parents of the patients, he reconstructed all patient histories. All of this costs a great deal of time, energy, and ultimately, money.

Dr. Buras and many other New Orleans physicians were impacted by the loss of charts and medical records in the aftermath of Hurricane Katrina. They failed to back up electronic data and/or make copies of paper charts. Water completely destroyed financial and clinical information, forcing many to close their practices and accept positions in other communities.

One of us, NHB, a physician whose practice survived Katrina, knows first hand the importance of backing up data. By having a practice management system and medical records electronically preserved, NHB was able to manage the practice while being evacuated for several months; unfortunately, however, hundreds of colleagues were not so lucky, as they lost all of their practice management data as well as their paper charts. With this loss, many closed their practices and accepted positions in other communities. This chapter, written after Katrina's learning experience, serves as a warning to health care professionals about the risk of not having a bulletproof data backup system. In this age of technology and Internet accessibility, it is possible to feel secure about your data with state-of-the-art online backup services, which automate the process of protecting electronic data. After implementing the information learned in this chapter, you will be able to operate your practice with just a few key strokes on your computer.

KEY QUESTIONS

1. Is your data backed up daily, hourly, or every time an entry is made into your practice management system or your electronic medical record?
2. Is your data backed up and stored onsite, offsite, or over the Internet?
3. Have you recently tested your backup program, and are you certain that the data can be restored to any computer in the event of a disaster?
4. Are your passwords changed on a regular basis?

Most medical practices and most other non health care professionals do not consistently and reliably back up their data. Most practices leave their tapes in the backup drive on their computers and never verify whether the data are saved or confirm that the data can be reliably restored. Usually practices that try to restore data from their backup tapes are not successful. According to the Department of Labor Statistics, 40% of small businesses never reopen business after they lose their data.[2] Now it is possible to outsource data backup and use the Internet as the vehicle to secure important and valuable data. As can be seen from the events of September 11, 2001, it is wise to store the data offsite. Natural disasters or loss of data cannot be prevented; however, there are definitely ways to protect your data. This chapter answers the following questions:

1. What data should your practice protect?
2. How can your practice back up its important data?
3. What is the cost of backing up your data?
4. Which individuals inside and outside of your practice can help you design a data backup system?
5. What data should be protected?

Interestingly, in the health care profession, almost everything is, or soon will be, digital, including not only practice management systems, but also electronic medical records (EMR) data, e-mail correspondence, and now X-rays and pathology reports—the paperless office will soon become a reality for nearly every medical practice. Also, the Health Insurance Portability and Accountability Act (HIPAA) mandates that efforts are made to protect your data and that disaster-recovery processes are implemented. In addition to natural disasters such as hurricanes, floods, fires, and earthquakes, data theft must be prevented. Large companies and even the government have lost data from data theft. The Veterans Association recently experienced this when thousands of veteran patients had their identification data stolen.[3]

Today, it is possible to have virtual private networks that ensure that the data from your practice can be accessed from remote sites. A virtual private network is defined as a means for using the Internet to provide private, secure access to practice management systems or your electronic medical record applications to remote employees, partners, and even third-party payors. This ensures ready access to your data from remote sites. The benefit of this technology is that you can provide access to limited areas of the program such as insurance information, patient demographics, patient's clinical records, or all of these.

Medical practices are very dependent on digital technology and are at risk for losing these precious data. It does not take a computer expert to realize that every hard drive will eventually fail. Computers have moving parts, and all will eventually crash. With today's high-tech world of sneaky spyware and venomous viruses, you are in more danger of data loss than ever before. Computer viruses grew by as much as 11% during 2003 alone. In 1991, there were 300 computer viruses in the world. In 2005, there were about 140,000.[4]

Like tires on your car, the electronic circuits that your computer rides on will eventually wear down and blow out. When this happens, you can either grieve at your loss or simply restore your data with data backup software. Companies build computers and create new software with the concept of obsolescence. The question is not *if* the hard drive will fail but instead *when*. Also, computers are at risk for theft, and laptops are frequently lost or stolen. Seamless backup systems are also important because backup tapes and CDs are so easily stolen. Not only is the hardware at risk for theft, but malicious, disgruntled, or even careless employees can erase data. Human error, in which an employee accidentally deletes or overwrites data on the hard drive or on the backup tapes, instead of theft, is the most common cause of loss of data. A plastic surgeon in Beverly Hills, California, had all of his before and after patient photographs on his computer. These valuable photographs were thought to be backed up to his tape drive. A computer technician was installing a voice-recognition software program when the digital photographs were accidentally deleted. When an attempt was made to use the backup tapes to restore these priceless images, the data were missing. Fortunately, the plastic surgeon was able to hire a "hard disk doctor," who was able to glean many of the photographs from the hard drive. This harrowing experience created angst and required a very costly restoration. Now a nearly foolproof backup system is in place.

HIPAA mandated that large practices (more than four doctors) had to back up protected health information (PHI) by the year 2003.[5] The deadline for all practices, regardless of size, has been extended. Doctors can anticipate that the government will soon be mandating that practices back up PHI. At this time, there has not been any case law on backing up data. Doctors should understand that those systems that have the tape in the same building as the computer will not comply with HIPAA's mandate. Also, a tape that is taken offsite must be encrypted so that if it is lost or stolen nobody can have access to patients' protected health care data.

Regardless of how your practice is organized and the degree to which you rely on electronic technology, you depend on three types of information. First, your practice management system contains patient demographic and insurance information that you use to schedule patients, generate insurance claims, and follow up on collections. Second, your practice management system contains financial information that is used to generate regular reports on practice activity. Third, your clinical information on patients resides either on paper charts and/or in an electronic health record system. Depending on your specialty, that clinical information might include not only office visit notes but also laboratory, imaging, and other test results. Your data backup plan should secure all three types of data.

PATIENT DEMOGRAPHIC AND INSURANCE INFORMATION

You work hard to gain patients and provide outstanding medical care. Depending on the type of practice management system in use, you likely have a rich database of information on all of your patients. You can sort the information by location (i.e., zip code, age, gender, chronic conditions, insurance carrier, and medication allergies). This valuable information can be sorted into various categories and then backed up electronically or on paper. Thus, if disaster strikes, you can quickly identify currently scheduled patients, patients who are scheduled for surgery or other procedures, patients who have chronic health problems, patients who have medication allergies, patients who need timely follow-up, patients that require medication given in the office, and other patient groupings. Even if your office is closed, you can focus on those who need your help the most.

PRACTICE FINANCIAL INFORMATION

Information on the financial status of your practice resides in two places: some on your practice management system and some with your accountant, who prepares monthly and annual reports. If your practice is interrupted, you will want to be able to retrieve information to show your insurance agency and/or banker, as this will be important in obtaining business interruption insurance. Keep electronic and/or paper records for 2 to 3 years so that you will be able to demonstrate accurately your income and expenses before disaster struck.

CLINICAL INFORMATION

Finally, your patients' clinical information—records of previous visits, medication allergies, and treatment plans—will need to be accessed in an emergency. After all, patients' medical problems do not stop when a disaster occurs. Immediate access to their clinical information is important when they return to your practice or when they are leaving the area and see another physician or enter another practice system.

HOW CAN YOUR PRACTICE BACK UP IMPORTANT DATA?

When implementing the administrative, financial, and clinical systems in your practice, create a backup system from the outset. Your data can reside on a server(s) in either your office or elsewhere. If you have selected the client-site server option, you have the data that should be backed up to a second site on a regular basis. Just having data backed up to your in-office server will not be sufficient if they become inaccessible or are damaged from the disaster. If your office server is damaged, a backup at a remote site is necessary in order to restore the data. If you have selected the application service provider option, the data reside on a server that you rent from your vendor; thus, the vendor should be backing up the data on a regular basis.

Whether your data reside at your office or are on an offsite server from which you rent space, make sure that your backup system is entirely automatic, meaning that there are no tapes or CDs that you need to purchase and manually handle. There is no setup cost because everything is done online. The backup must be both safe and secure. It is safe because there are two layers of encryption, meaning that there are two backup servers at a remote distance. It is secure because the data are encrypted from the moment that they leave your computer. The data backup service does not have a copy of the encryption key—only you do. As the data had already been encrypted after they were transmitted over the Internet to the backup server, software on your computer creates a Secure Sockets Layer (SSL) encrypted connection that is the same as if you visit an online banking website or purchase something from Amazon.com using your credit card number.

The online backup of your data is less complicated than you might think. Your vendor(s) can help you take the necessary steps. First, register for a backup website, and select a password. Then configure your account and computers. Before actually backing up any data, create an encryption key, and select those files that you want to back up. Next, select the time of day that you want the automatic backup to occur. Most practices choose a time that is after office hours so that there is no disruption of the work flow.

When selecting a backup system, confirm that the system is HIPAA compliant. The backup company and its employees should not have access to your practice's PHI. Because there are two layers of encryption and also two servers, each of which has advanced firewall suppression systems and a 24-hour security monitoring device with biometric access or security consisting of fingerprints or facial recognition, you can rest assured that your data are well protected. With biometric access, only employees who have authorized access can log on to the computer system. Fingerprints or facial recognition are commonly used techniques. The backup servers have 24-hour battery and generator power backup so that a power outage at the site of either server does not damage or destroy your data.

WHAT IS THE COST OF BACKING UP YOUR DATA?

The cost of backing up your data depends on the amount of data being stored. Plans start at $10 per month and may cost as much as $200 per month. The average cost of data backup for a small practice is $25 to $50 per month. Most companies bill the practice for only compressed data; therefore, verify that the company is not billing you for total raw data size but only for the compressed size of your data.

HOW TO SELECT THE CORRECT DATA PROTECTION SOLUTION

The best resources are those information technology companies that have experience in working with data protection for medical practices. Important questions to ask the vendor are as follows: How much data can we afford to lose if our computers went down? What is the longest electrical outage that we could tolerate? You must place a monetary value on the data and determine their importance to the practice if unavailable. After these questions are answered and a determination is made, then an assessment of the value versus the cost in protecting that data can be determined, and the best data protection system can be implemented at the most reasonable price.

At a very basic level, data can be protected by making copies of the data. The method that is used to make these copies determines the level at which the data are protected as well as the costs required to reach that level. How often are those copies made? Thus, if data are captured after the last copy is made, that additional data will be lost if a disruption or disaster occurs. For example, if your program backs up data each evening but an electrical outage occurred in the afternoon, all of the data that were entered in the morning and early afternoon would be lost.

TYPES OF DATA BACKUP

Tape backup provides the lowest level of protection and results in the greatest potential data loss and the longest time to restore in the event of a disruption. Many small- and medium-sized practices use tape backup for both disaster and operational recovery, and larger practices use a tape for only a portion of their applications. As more practices put larger amounts of data online, every practice has a challenge to execute these backup procedures in a failsafe manner. Also, as the volume of data increases, the backup time becomes increasingly longer. This usually means that the backup takes place when the practice closes or in the middle of the night. Thus, any data entered after the last backup will not be captured if a disaster occurred between backup times. Many practices use high-capacity, low-cost disks for backup. They are faster and more reliable and provide higher protection levels than slightly more expensive tape backups.

Doctors are also deploying point-in-time copies. These copies can be a snapshot or a clone of your data. A snapshot is not an actual copy of data but a pointer to the original data as the data were when the snapshot was taken. A clone is a separate physical copy of data created at a specific point in time. Snapshot technology allows a copy of your data to be made with much greater frequency because a snapshot is instantly accessible, unlike a total copy of data, which must be synchronized before it can be accessed. A snapshot also consumes less disk space than a clone. The primary benefit that snapshots or clones provide over disk or virtual tape is the ability to make copies more frequently, which decreases the recovery time objective.

Continuous data protection is a new technology that is designed to capture data modifications continuously and store changes independently of the primary data, enabling recovery points from any point in the past. A major advantage of this is that it preserves a record of every transaction. Continuous data protection can protect against data corruption errors because data can be rolled back to the exact instant before the error occurred.

WHICH INDIVIDUALS INSIDE AND OUTSIDE OF YOUR PRACTICE CAN PROVIDE ASSISTANCE FOR A DATA BACKUP SYSTEM?

Several resources can be used to locate a data backup system for your practice. An information technology (IT) expert or a practice management software vendor may be able to make a recommendation. Finally, if you are using an electronic medical record that uses a local server and also an Internet backup at a remote site, you can feel secure that your patients' records are safely backed up. This latter program is often referred to as the blended model. With the blended model, data are also saved on a master server in a remote location that is updated on a real-time basis over the Internet. With this program in place, if the local server is down, then the practice's data goes to the Internet and logs on to the master server until the local server is repaired. As a result, you do not risk losing data if the local server is malfunctioning or if the Internet is down—the data can still be preserved on the local server. This is one of the most secure methods of saving and backing up your data.

The selection of a backup system is going to be based on the technical requirements of your practice. You should strongly consider an online backup system for your practice. Why? It is entirely automatic. No hardware is involved, and no tapes or CDs are required. It is safe and secure, as it is done offsite. There are no upfront costs, and the system is completely scalable (i.e., there are no setup fees, which is unlike traditional backup methods [i.e., tape] that require expensive hardware and supplies). If you are using a tape backup system, good-quality tapes cost $50 to $100, and the hardware for a first-class tape drive is approximately $1,000. Additionally, tape drives support only one type of backup tape (i.e., digital linear tape [DLT] or linear tape-open [LTO]). These tapes have limits on how much data they can store. If your data volume grows too large—especially if you are backing up digital images from a radiology or pathology practice, for example—you will need to buy a new tape drive and throw out all of your existing tapes. Online backup can keep expanding as your data continue to increase. For example, many hospitals are converting to picture archiving and communication system images. Each image can be several gigabytes in size. Backing up and restoring these images on tape or on an optical disk could take hours. Now this can be done in minutes and is less expensive if done online.

Using an online backup system is very cost-effective. There are no upfront costs, and the program is scalable and never obsolete. An online program requires no staff involvement and is entirely automatic. This is compared with a tape backup or restoration, which can take hours, especially if the practice has large volumes of data. Online backups and restorations are instantaneous. The transfer of large amounts of data is limited by only the speed of the Internet connection.

Two other major advantages of online backup are safety and security. Online backup is safe because your data are always stored off-site and are encrypted from the moment the data leaves your computer. A safe backup system will uses two layers of encryption—that is, data are sent to one location and then to another server, usually at a great distance from the primary storage site. Thus, if there is a disaster that encompasses large geographic areas, the data will still be secure because they are stored at a second site. For example, a practice in New York City uses an online backup program located in northern New Jersey, but with a secondary storage in Denver, Colorado, the data will be secure if there is a shutdown affecting both the New Jersey and New York City areas. By using two layers of encryption, the data are encrypted using the customer's unique 256-bit AES (advanced encryption standard) secure encryption key that only the customer has. Because the data backup service does not have a copy of that encryption key, no one is able to access the practice's data. Second, as the data (already encrypted once) are being transmitted over the Internet to the server, the software on the client's computer creates an SSL connection to our servers. An SSL connection is an encrypted connection that is the same as if you visit an online banking site or buy something from Amazon.com.

Only the person(s) who has the encryption key has the ability to restore the data. The data that are stolen from the servers in your office or at the backup site would be useless to the thieves.

Getting started with online backup is very simple. You can register and sign up on a website. It takes 10 to 15 minutes to install and configure your account and computers. Before backing up data, you must create an encryption key and then select the files to be backed up. Then select the time of day that the back up is to occur, usually after the office closes. Using this technique, you do not tie up the computers that are used for daily operations.

One online backup vendor, eSureIT, from Intronis Technologies, is a secure online backup service that automates the process of backing up electronic data. eSureIT was created to satisfy the broad need for an easy-to-use, automatic, and secure method of backing up data offsite. The goal of eSureIT was to design a cost-effective backup service that could be used by any medical practice, regardless of the size and the computer expertise; nevertheless, it needed the functionality and features of backup systems used by Fortune 500 companies. Since its market introduction, eSureIT has quickly gained recognition with customers nationwide; these customers have come to recognize that eSureIT provides a backup solution that is simple to setup, easy to use, completely automatic, and most importantly, secure and reliable.

After signing up for an eSureIT online backup service account, the practice administrator or IT person will download and install the software onto his or her computer or server or onto multiple computers or servers. After the software is installed, the user will be prompted to choose a unique 32-character string of characters (the encryption key) that will be used to encrypt all of the user's files. This encryption key is stored on only the user's system and is never transmitted over the Internet, nor is it stored on Intronis's servers. Thus, only the user has access to his or her files. Even Intronis' employees cannot access the files or even read the filenames.

Next, the administrator or user will create a backup set, which is the list of files to be backed up and the days and times that backups will run. Backup sets can be created by selecting either individual files or folders or the types of files to backup. In addition, multiple backup sets can be created, allowing the user to have complete control over the data being backed up. The first time a backup occurs after a new backup set is created (the "initial backup"), all files contained in the backup set will be transmitted. Thereafter, only new files and files that have changed will be uploaded (an "incremental backup"), minimizing the time that it takes to perform a backup and the user's valuable bandwidth. Backups will begin automatically according to each backup set's schedule, as long as the computer is on and functioning (and not in sleep or power-save mode). The user can initiate backups at any time. Because backups run in the background of the system, they have little or no impact on the computer's performance or Internet connectivity.

When a backup starts, the system's hard drive is first scanned for any files contained in the backup set that are new or have changed since the last backup. eSureIT allows the user to store an unlimited number of versions of a file (the default setting is 30 revisions per file). Storing multiple versions ("revisions") of files is useful when the content of files, but not the filename, changes often. The danger of accidentally overwriting a file is thus eliminated.

After the eSureIT software identifies a file that needs to be backed up, it compresses the file using industry-leading compression (ZIP) technology. Compression ensures that not only do backups take a shorter period of time but that the amount of storage space used is minimized.

After being compressed, each file is individually encrypted using the unique 256-bit encryption key. eSureIT uses the 256-bit AES encryption technology. AES encryption was developed by the U.S. National Institute of Standards and Technology and is now the state-of-the-art standard encryption technique for both commercial and government applications. Moreover, in June 2003, 256-AES was approved by the U.S. National Security Agency for use in encrypting the U.S. government's "top-secret" documents.

For added security, each encrypted file is then sent over the Internet via a secure channel using SSL technology. This same technology is used for online banking and online credit card applications. Thus, data are encrypted twice. They are encrypted at all times using the 256-bit AES encryption again while being sent over the Internet to and from the Intronis servers.

All user data are sent to and stored in two redundant secure data centers that are located hundreds of miles apart from each other (Northern New Jersey and Toronto, Ontario, Canada). Each data center has 24/7 onsite monitoring, advanced security technology such as biometric access controls, backup generators, and redundant connections to the Internet. Each file that is backed up or restored, as well as additional information and statistics about backups, is recorded in a log within the eSureIT software. This log, which can easily be searched, allows the user to verify that files were successfully backed up and helps to troubleshoot ongoing issues. The user also has the option of receiving an automated e-mail notification at the conclusion of each successful backup. Information about recent backups and total storage use can also be viewed on the Internet by logging on to the user's account at www.intronis.com.

Of course, any backup method is only as good as its ability to retrieve files. With eSureIT, restoring files can be done in just a few clicks of the mouse. With the eSureIT software, the user simply clicks on the individual files or folders or revisions that he or she wants to retrieve. The file(s) will then be downloaded to the user's computer, decrypted, uncompressed, and then restored to their original location or another specified location on the user's system. A password is required to restore any files, preventing unauthorized restores.

In the event of a complete system failure, a full recovery of the user's backed up data can be initiated in minutes. This recovery can be done on any Windows-based computer, not just the computer from which the files were originally backed up. The user simply downloads and reinstalls the eSureIT software, enters his or her username and password, and loads or types the encryption key. After the software installation is completed, two clicks of the mouse will restore the file catalogue or backup sets (the list of all of the files backed up), which will then give the user the ability to restore any and all files that are backed up.

The backup system is HIPAA compliant, and the backup company and its employees do not have access to the PHI. In addition to the two layers of encryption, there are two servers, both with advanced firewall suppression systems and a 24-hour security monitoring service using biometric access, which means that only those employees who have authorized access using their fingerprint or facial recognition can log on to the computer system. The program's two servers have a 24-hour battery and generator power backup, which ensures that a power outage at the site of either server will not result in the loss of your data. (Other resources for Internet backup are included at the end of the chapter.)

TechAtlas (www.techatlas.org), which allows you to have a free user account, is the quickest and easiest way to inventory your in-office computers and the electronics. If you use this program, which you can install on your computer, you will be able to automatically record all of the information you need about your computers, servers, and software programs, including the model, serial number, and other details about your hardware and software. You can upload the result to TechAtlas, which creates a spreadsheet for you.

All of the equipment that is used in the practice is included in your technology inventory form. This includes the model number, serial number, and information about components and customizations for your computers, fax machines, copiers, and so forth (Figure 6-1).

For insurance purposes, it is useful to have proof that you had the items. You will also need photographs of your entire office to record all of your furniture and clinical equipment, including laboratory equipment, X-ray equipment, EKG machines, and so forth. These photos, which should be updated at least annually, can be burned to a CD and added to the disaster box that was described in Chapter 3.

Consider having a backup of your software operating system, office applications, database programs, antivirus programs, and so forth. This will make it much easier and quicker to set up replacement computers if the existing ones are destroyed. It is legal to burn a backup copy of all software, and many CD copying programs are available online for under $30.

IT ISN'T ENOUGH TO HAVE A PLAN— YOU MUST TEST THAT PLAN

More than one practice thought that they had a backup system in place; however, when the computers crashed or were nonfunctional, these practices found out that the backup system was ineffective and that the data were lost. Because this is such a common occurrence, we are recommending a disaster recovery "fire drill." This means giving the responsibility to someone in the office or bringing in an information technology person at least every 3 to 4 months. To have a drill, pretend that your server hard drives have crashed and that you

Chapter 6: Creating a Backup Plan

DISASTER PREPAREDNESS	*Computer Workstation Inventory*
CONTINUITY OF OPERATIONS PLAN (COOP)	Updated on: _____ by:_____

General Information

User	Department
Vendor	Purchase date
Brand	Model
Model #	Serial #

Hardware Specs

Processor (mHz)	Hard drive (GB)
Memory/RAM	CD drive
Other hardware	

Software/Application

Operating system	Office version
Antivirus version	
Other software	

Monitor

Brand	Model
Model #	Serial #
Vendor	Purchase date and price

Notes

Figure 6–1 Computer workstation inventory.
Source: Louisiana Association of Nonprofit Associations © 2007.

need to restore the data. Now the information technology person has the responsibility of proving that the backups are viable. This drill also will give you a feeling of security and will let you know how long it will take to get back up and running. If the person responsible for backing up the data is unable to reproduce the data, then there are problems with the disaster recovery plan. The person in charge can then tweak the plan to produce a better result. You must have a data recovery plan and test that backup plan, especially the data that are essential for getting your practice operational after a disaster. If you regularly check your office's backup ability, you will be left with little down times and little to no loss of vital data.

BOTTOM LINE

Backup—superheroes need it, police rely on it, and everyone that uses a computer should use it. In the world of mainframes and microchips, it is called data backup or data recovery, and it can mean the difference between a slight computer setback and living through your own electronic apocalypse, a digital Pearl Harbor, or a digital 911!

Our computers are a bigger part of life than ever before. We shop, work, and play using computers. They have replaced stereos, dictionaries, encyclopedias, and even the mailman. Computers and the data stored on them are fallible—just like doctors, nurses, and hospitals; therefore, we must be sure that data are protected, backed up, and stored in a secure place in the event that a disaster impacts our practice. The Gulf Coast and Eastern Seaboard are not the only places that are impacted by disasters. Data loss can occur in *any* practice. Be sure to back up your system; then check and restore a file periodically, thus verifying that the system is in working order. Remember that *anything* you do is better than doing *nothing*!

OTHER RESOURCES FOR BACKING UP YOUR DATA

1. RBackup Remote Backup Software (http://remote-backup.com/) works like regular data backup software, but with one important difference: Instead of sending backups to a tape drive or other media, RBackup online backup software sends the backup over the Internet, regular telephone lines, or other network connections to your offsite online backup server. It

typically does this at night while computers are not being used. Backups can also be done on demand, any time. It is completely automatic. In fact, you may even forget that it is working. Most businesses put their lives on the line every night and do not realize it. With businesses depending more and more on the data stored in their computers, proper backups are becoming much more critical.

2. See http://www.novastor.com/.
3. See the online article "All Veterans at Risk of ID Theft after Data Heist: Burglar Reportedly Took Veterans Affairs Disk Containing Personal Info," (http://www.msnbc.msn.com/id/12916803/).
4. A TK8 backup (http://www.tk8.com/backup.asp) is reliable, easy-to-use backup software that ensures that you never lose important files. It protects your data from loss (e.g., file corruptions) caused by computer crashes, hard disc failures, human errors, theft, virus attacks, and so forth.
5. See http://www.stompsoft.com/pc-backup/.
6. See http://www.maxtor.com/. This program offers one terabyte of capacity, or 1000 gigabytes, for shared storage and automatic back up of networked PCs and Macs.
7. See http://data-backup-software-review.toptenreviews.com/why-backup-your-computer.html. This site provides a comprehensive comparison of several software vendors and a head-to-head comparison of the specific specifications and unique features of the various programs.

References

1. Intronis Technologies. "About the eSureIT Online Backup Service." Available at: http://www.intronis.com/esureit.html, accessed February 11, 2008.
2. D'Amico, V. "10 Steps for Surviving a Disaster!" *Handbook of Business Strategy*, 2004.
3. Statement of Steve Robertson, Director, National Legislative Commission, The American Legion. February 12, 2007. Available at: http://veterans.senate.gov/public/index.cfm?pageid=16&release_id=10776&sub_release_id=11051&view=all, accessed February 11, 2008.

4. Hammond, S. "Busting the botnet-herders, interview with virus expert Mikko Hyppönen. November 1, 2005. Available at: http://www.techworld.com/security/features/index.cfm?featureid=1926&pagtype=samecat samechan, accessed February 11, 2008.
5. U.S. Department of Health and Human Services, Office for Civil Rights. "Summary of the HIPAA Privacy Rule." May 2003.

Chapter 7

Before and After a Disaster: A Hospital Perspective

In preparation for battle, I have always found that plans are useless, but planning is indispensable.

—Dwight D. Eisenhower

Imagine being in the direct path of a Category 5 hurricane (winds greater than 150 mph) and having the responsibility for a major hospital. Indeed, Hal Leftwich, DBA, CEO of Hancock Medical Center in Bay St. Louis, MS, faced this scenario before Hurricane Katrina hit on Monday, August 29, 2005. Although the prestorm planning, using "Coding White" (impending hurricane), was excellent and began on Friday, August 26th, the hospital was in lockdown 36 hours before landfall; however, the 100 employees and 33 patients did not realize that within the next 36 hours the hospital would be inundated with 9 feet of water. Many noncritical personnel and physicians had left town; thus, only a handful of physicians remained to care for already hospitalized patients, as well as those who suffered injuries caused by this enormous storm. Because of the hurricane, the Emergency Operations Center (EOC) within Bay St. Louis was completely destroyed; therefore, the hospital could only communicate with certain members of the EOC for about 3 days. Immediately before landfall, all patients were moved to the second floor, and both the National Guard and Coast Guard were onsite on the afternoon of August 29, 2005. Within 48 hours after the storm hit, over 800 people were treated in the emergency department! Search and rescue operations began immediately on the afternoon of August 29th. These patients were sent to Hancock Medical Center; however, by Monday evening, all patients were transferred to South Alabama Medical Center in Mobile, Alabama, and the hospital was totally evacuated by Wednesday, August 31, 2005.

Surprisingly, although the hospital went without air conditioning for several days, the major generator was repaired for critical operations the afternoon of August 29th. Various disaster medical teams from Missouri, Florida, and Iowa were dispatched directly onsite and began arriving on Wednesday, August 31; these emergency teams continued to operate in town until Thanksgiving 2005.

William Tate, Chairman of the Board of Directors of Hancock Medical Center, was so instrumental in providing onsite leadership that he was cited as the 2006 Trustee of the Year by Modern Healthcare (small hospital category). Furthermore, several board members assisted in the aftermath of the storm even though six of the seven of them lost their homes, as did the majority of hospital employees and physicians.

According to Mr. Leftwich, disaster planning for hospitals means that they should be self-sufficient for about 3 days; however, given the enormity of this disaster, the new thinking is that hospitals should be self-sufficient for 7 to 10 days, particularly with respect to drinking water and water needed for other sanitary purposes. In addition, because communications were most severely impacted, Mr. Leftwich recommended that hospitals establish a HAM radio setup with a domed antenna, as this appears to be an effective method of communication during a catastrophic event. Although some hospitals have satellite telephones, many do not have a sufficient supply for key personnel because of the high expense. Also, high-powered walkie-talkies appear to be extremely useful for interagency communication up to 6 to 8 miles. Finally, Mr. Leftwich indicated that vendor support is very critical; however, few companies have disaster plans and thus fail to make accommodations for equipment and supplies.

Touro Infirmary in New Orleans was one of the few hospitals within the Metropolitan New Orleans Area that was able to operate immediately after the impact of Hurricane Katrina.

Mr. Leslie Hirsch, CEO of Touro Infirmary, like Mr. Leftwich, realized that hospitals cannot operate without a medical staff. Because many physicians had damage to both their homes and offices, Touro Infirmary made accommodations for physicians through a rent-abatement plan for those who were renting medical office space from the hospital; it also offered the use of emergency communications through both the hospital and their websites.

Key Questions

1. Does your hospital have an established disaster plan that includes the physicians on the medical staff?
2. What is expected of you and your employees if a regional disaster impacts your community?
3. How will your hospitalized patients be cared for if a disaster requires evacuation from the hospital?
4. Are you aware of what services or assistance your hospital could provide if a disaster impacted the community and the building where you practice?

Hospitals should be prepared to support their medical staff in every possible way—from medical office space to potential physician employment (short or long term), information technology support, and financial assistance. Indeed, the board of directors of Touro Infirmary authorized 5 million dollars for medical staff recovery! This money assisted doctors in returning to practice and also subsidized those with diminished incomes because of the impact of this disaster. Although this type of assistance may be inconsistent with Stark Laws, the government relaxed these requirements during the aftermath of this particular disaster.

Mr. Hirsch stated that each medical practice should have a detailed disaster plan that coincides with the hospital's plan, particularly with respect to communication with the hospital via e-mail. Furthermore, the hospital also needs to establish an emergency website and advertisements in major newspapers within a 200-mile radius of the disaster. They also attempted, via a call center, to locate and update physicians on the status of hospital operations.

Indeed, each medical practice should coordinate its key contact information with the hospital, and likewise, the hospital should supply each medical practice with emergency communication numbers, websites, e-mail address, and so forth.

BEFORE A DISASTER

According to Frank Folino, vice-president and director of emergency planning at Touro Infirmary in New Orleans, if you ever anticipate an impending crisis, you can execute a disaster plan at the hospital level. Before Katrina in August 2005, a very limited plan was in place to manage the devastation of a direct hit of a category 4 or 5 hurricane. Now a disaster plan has been written and is ready to implement on very short notice. For example, approximately 120 hours before a hurricane impacting the Gulf Coast region, disaster-planning process is started.

At Touro Infirmary, a Hospital Emergency Incident Command System (which consists of a medical staff director, a group of physicians who have been selected and agree to be a part of the disaster recovery, and the recovery team, which consists of most medical specialties that are involved in trauma and managing most medical emergencies) is in place.

The hospital's plan requests that all of the doctors report their evacuation plans to the medical director. The disaster plan needs to specify which doctors are going to stay in the area and how they can be contacted. It also needs to contain contact information, including locations, phone numbers, e-mail addresses, and physical addresses for those physicians who are leaving. At Touro Infirmary, the doctors are given appropriate credentials that identify them as members of the medical staff, as this allows them to easily reenter the city after the disaster. All doctors should take the credentials with them before leaving the area. After the disaster plan has gone into action, the team will let the doctors know the status of their patients as well as the status the hospital; it also informs the physicians about what additional help is needed.

The team must be able to manage and direct the care of each patient who remains in the hospital. When the disaster plan becomes active, all elective surgery is canceled, and the operating room is prepared for only emergency surgery.

The emergency team directs the medical care and disposition of each patient in the hospital. It will ultimately make decisions about which patients are to stay in the hospital and are unable to be moved, which patients can be discharged, and which patients can be evacuated to regional hospitals that are accepting patients from the anticipated storm area. The objective of the disaster team is to reduce the census. The physicians on the disaster team are ultimately responsible for those patients in the hospital during and immediately after the storm.

After the disaster plan is activated, the doctors on the medical staff can communicate with the hospital via multiple methods. For example, Touro Infirmary has a website that contains a link to information about the hurricane. The website allows the hospital to post messages to physicians on the medical staff, as well as to the employees of the hospital. It also provides a way for patients' family members to view the status of the hospital and gives them numbers to call regarding the status of their family members who remain in the hospital. The website provides a toll-free number that physicians and family members can use to contact the hospital.

After the disaster has occurred, a hospital's disaster plan provides the steps for recovery. Touro Infirmary has a recovery team that goes into action 3 to 4 days after the storm has passed. In a widespread disaster such as a hurricane, the National Guard is usually deployed to protect the city and its institutions, including hospitals. Doctors will need to know the reentry plans so that they can return to the city and the hospital. They also will need to have appropriate credentialing in order to pass the National Guard check points. At Touro Infirmary, the physicians on the medical staff and first responders have placards provided by the city that allow them to return to the city and the hospital, as well as through the checkpoints set up by the National Guard. Finally, Mr. Folino suggests that physicians on the medical staff keep in regular contact with the hospital. Medical staff should also inform the hospital when they are returning to practice.

WHAT CAN YOUR HOSPITAL DO FOR YOU AND YOUR PRACTICE?

If your practice is destroyed, the hospital can often help to find temporary locations for your practice and can sometimes offer financial assistance. It can help provide the temporary supplies, such as medications, that are necessary to start your practice. The hospital can be invaluable in providing security for your practice.

Because many of the businesses in the community may not be open for weeks after a disaster, your hospital may be helpful in providing nonmedical services such as loans, laundry, food, and a post office to make your practice operable.

On May 27, 2007, the New Orleans *Times-Picayune* ran an extensive article entitled "Hospitals to Operate Under Storm Rules."[1] According to Staff Writer Kate Moran, most local hospitals plan to stay open during a major hurricane with a number of key changes, such as establishing secure command centers with satellite telephones, elevating emergency generators from ground level, as well as making evacuation plans for newborns and intensive care unit patients days before the impact of a major hurricane. Furthermore, virtually all hospitals have adopted the new philosophy of the stocking of water and supplies for 7 to 10 days, as well as making prestorm planning arrangements for patient transfers with regional hospitals. Furthermore, in addition to enhanced communication capabilities, most hospitals are also establishing telephone and Internet access to assist in employee and physician communication. Interestingly, many people stayed after the hurricane because evacuation units were reluctant to accept family pets; Tulane University Hospital and Clinic, however, has established secured kennels to house pets of staff members who are prepositioned at the hospital during an emergency.

BOTTOM LINE

Because of the mutual dependency between physicians and hospitals, hospitals must take the lead in disaster planning and hold biannual meetings with key staff personnel to exchange critical contact/communication information in the event of a natural disaster. Your hospital can be your ally before, during, and after the storm. The key is keeping avenues of communication open.

REFERENCE

1. Moran, K. "Hospitals to Operate Under Storm Rules," New Orleans *Times-Picayune*. May 27, 2007.

Chapter 8

Insurance for Ameliorating the Pain of a Disaster

There are worse things in life than death. Have you ever spent an evening with an insurance salesman?

—Woody Allen, U.S. movie actor, comedian, and director

A fire destroyed both the office and the building of a physician from the Pacific Northwest. Although he had insurance for the property, medical equipment, and furnishings, he did not have business interruption insurance. As a result, he was unable to support his staff and had to use his personal savings during the rebuilding of his office. The project took several months to complete, but the financial hardship could have been completely avoided if the physician had purchased business interruption insurance.

Key Questions

1. Have you met with your insurance agent within the past year to review your insurance needs?
2. Do you have business interruption insurance?
3. Do you have disaster insurance?
4. Is a copy of your insurance policies kept offsite?

Hopefully this book has convinced you that disasters are not rare or unlikely. A disaster is a possible occurrence for many medical practices. If it occurs, weeks and even months may pass before the practice is again fully functioning. Damage may occur to your building and to the property inside, and ongoing expenses must still be paid. This chapter suggests insurance options for your practice to reduce the financial impact of unforeseen events.

According to the Institute for Business & Home Safety, a non-profit insurance-industry organization, at least one quarter of all businesses that close because of a disaster never reopen. The toll is even worse for small businesses and small medical practices. Federal experts say that 40% of small businesses that close for a major disaster (such as Hurricane Katrina) are closed for good.

This book provides plenty of advice about how to prepare your business for disaster, but this chapter specifically discusses the insurance and contingency planning that is necessary to reduce the impact of unforeseen events.

SELECTING DISASTER INSURANCE

When choosing disaster insurance, two choices are available: a peril policy or an all-risk policy, also known as a comprehensive policy or an open peril policy. The primary difference between the two is that one type of policy covers what is listed in the policy, whereas the other covers what is not.

A named peril policy is often a good choice for those practices that are located in areas that are frequently hit by natural disasters such as hurricanes, tornados, or floods. Such a policy spells out the specific events for which you are covered. The cost of the premiums depends on the location of the business and the likelihood of the specific peril(s). Anything not specifically named in such a policy is not covered.

An all-risk policy protects your business from damages caused by any type of disaster, with the exception of those specifically excluded. Floods and earthquakes are two events that are typically excluded, but coverage for these types of disasters can be added for an additional fee. The National Flood Insurance Program (NFIP) underwrites coverage for flooding, making it more easily available to medical practices. (Check the informative government website www.floodsmart.gov.) Flood insurance policies are continuous—they are not canceled or renewed for repeat losses. Flood insurance reimburses medical practices for all covered losses up to $500,000. The average cost of a $100,000 flood policy is a little more than $400 annually, or just over $1 per day.

For most businesses, an all-risk policy (with perhaps a rider to cover flooding) will suffice; however, only you can determine your needs based on your geographic location and the property and equipment that you want to protect. The advantage of an all-risk policy is that it covers you in the event of an unpredicted disaster, and many unusual disastrous events fall into this category. For example, one medical practice in a rural town was pleased that it chose an all-risk policy after a horse pulling a carriage became startled and backed up quickly, sending the carriage through the doctor's front window and into the middle of his reception area. There was no exclusion for horse-drawn carriages coming through the window, and therefore, the practice was able to collect.

We suggest that automatic coverage is included for contents at a percentage of the value of the building. Currently, building contents are not covered in an NFIP policy unless the insured specifically decides to include that coverage and cost in their NFIP policy. Most all in the NFIP program have building coverage, but far fewer have their contents covered; however, when a loss ensues, doctors forget and confuse the manner in which property insurance covers a loss when the building and the contents are damaged. Many expect that contents will be covered as a percentage of the building value covered. Businesses can be insured for up to $500,000 in building coverage and up to $500,000 in contents coverage. Building insurance must be supplemented with an extra policy for your contents.

Much of the damage from Katrina was caused not by wind, but by the storm surge and the failure of the levee system in New Orleans. Insurers define these events as flooding, and thus, property owners must look to flood insurance for coverage. Unfortunately, many medical practices in the Gulf Coast area had not purchased federal flood insurance and were thus not covered for the damage that the hurricane's storm surge caused.

Today, numerous choices and comprehensive insurance options exist—you can typically purchase a small business package that meets your needs; however, you must determine what you are insuring and for how much. Take stock of your medical property. Determine its value, and decide what is worth insuring.

PROPERTY INSURANCE

Property insurance covers damage to or loss of personal property or equipment in your practice. This is different from liability insurance, which covers claims from third parties (e.g., a patient who is injured when he or she trips on your torn carpet).

Review and update your insurance polices periodically. Annually evaluate all of your insurance for appropriateness of coverage, premium expense, and exclusions. Keep in mind that you can increase or limit your coverage over time. Often, practices simply let their insurance coverage carry over from year to year without adding coverage for recently purchased medical equipment or business-related items. This is one more reason to review your current coverage with your agent.

If you buy all of your insurance from one carrier and one agent, you can usually save money with riders and packages. One-stop purchasing means that your coverage is less likely to overlap. Before selecting a carrier, get some recommendations from other medical practices in your community. Are they are pleased with the service and the premiums? In addition, always ask questions about anything that is unclear in the policy.

Take the time to use a video camera to document the property within your office. Very few of us have a memory sufficient to withstand the rigors of a cross-examination about the "proof of loss" in an insurance dispute. Video tape the outside as well as the inside of the building, especially if you own the premises. Do not forget the landscaping, as that will need to be replaced that as well. No cliché rings more true than "a picture is worth a thousand words" when you are trying to explain your loss to an insurance adjuster.

To determine how much insurance to purchase, consider the replacement value of your equipment and furnishings. Many insurance policies cover only the actual cash value of your property, which is often the depreciated value of your property. On the other hand, if you are going to replace this property, you will probably want to insure it for what it will cost to replace it, and this kind of premium is more expensive than a policy that covers the actual cash value of your property. Ask your agent or insurance broker to explain the difference in cost between actual cash value and replacement value coverage. Then you can determine how you want to insure it.

BUSINESS INTERRUPTION INSURANCE

After Hurricane Katrina devastated the Gulf Coast area in 2005, hundreds of practices were shut down for weeks or months. The same happened to businesses and medical practices south of Manhattan after September 11, 2001. The message is this: Disasters do happen and can interrupt your ability to practice medicine. Although most interruptions are not on such a grand scale, these practice interruptions do occur. Your practice and ability to maintain your livelihood can be destroyed or temporarily rendered unable to function within an instant and without any warning.

Because many physicians' sole or major source of income depends on our office being open for business, they must consider the possibility that circumstances could compel us to close our practices.

Business interruption insurance is a necessity to make up that lost income. In essence, it protects a lost earnings' stream, with earnings defined as gross revenues minus expenses (including physicians' salaries). Business interruption insurance (also known as business income protection, profit protection, or out-of-business coverage) provides funds to make up the difference between normal income and income during a forced shutdown after a disaster.

Although property, liability, disability, and other types of insurance can provide medical practices with protection against specific risks, most policies do not cover the indirect costs associated with disasters. When a medical practice suffers a loss, as in the case of property damage in a fire, it may be forced to shut down for some time or move to a new location, either temporarily or permanently. A typical property damage policy will cover the cost to repair or replace buildings and equipment, but it will not cover the loss of income. The practice thus may be forced to tap cash reserves in order to pay continuing expenses—such as taxes, salaries, and loan payments—even when the practice has no income. In addition, practices may face extra expenses in a crisis, such as employee overtime or rent on a temporary location. Finally, practices confronting temporary operational shutdowns are faced with dramatically curtailed revenue and the prospect that patients—whether long time or potentially new—may establish relationships with other physicians and other practices.

Business interruption insurance is usually sold in conjunction with property coverage at the same level of cost. For example, doctors who pay 15 cents per hundred dollars for the replacement value for their property insurance will usually pay a similar rate for their business interruption coverage.

Business interruption insurance is usually triggered by damage to the property where the business is conducted. There is usually a deductible in either a flat dollar amount or a waiting time. For example, if waiting time triggers the deductible, the policy may require at least 24 hours to pass after the business has been disrupted before the payments begin. Most business interruption forms do not include coverage for perils such as emergency evacuation by civil authority or a major utility disruption. If you want this coverage, it must be added with a rider.

Physicians need to consider the pros and cons of this kind of business interruption coverage. Doctors just starting out in a practice probably do not need it, as they have little or no earnings to protect. The amount of reimbursement under any interruption policy is directly based on the business's level of gross earnings; however, a physician who has been in practice for several years may need to protect his or her revenue stream and thus should be very interested in business interruption insurance. In fact, the more income the physician's business produces, the greater the need becomes, as there is so much more potential for loss. A high-income medical practice is a prime candidate for business interruption insurance.

When deciding about business interruption insurance, consider factors other than your level of income. What is the probability of a practice being affected by something that could shut it down? What are the predominant weather patterns in your region? The higher the probability that tornadoes or hurricanes will affect your practice, the more important it is for you to purchase business interruption insurance.

A big-city practice, however, has its own concerns and risks. Doctors whose offices are located in high-rise buildings should investigate whether the property has adequate sprinklers or other fire-retarding features, such as fire doors and walls. Building codes are not uniform throughout the country, and some structures have been granted compliance waivers even though regulations have tightened for new buildings. The presence or absence of these features can make the difference between minor damage from a quickly contained blaze versus widespread destruction that shut downs all of the businesses in the high-rise and triggers the need for income replacement.

How long would it take to resume operations? How much would it cost to put a disrupted business back into operation? Think about the possibility of relocating your practice either permanently or temporarily if a disaster occurs. Office space may become difficult to find, and even if found, the demand may be so great that the cost could be prohibitive. Also, consider other expenses such as notifying your existing patients of your new location.

In some respects, physicians are in better shape than a retail establishment if a short-term change of business location becomes necessary. If a retail store moves a few blocks down the street, it may lose customers to a more conveniently placed competitor. Patients are more likely to follow their physicians across town, especially in these days of managed care, when it is much harder to change designated primary providers; however, moves of more than 25 miles will find most practices with an erosion of their existing patient base.

Nonetheless, because of the importance of the direct physician–patient relationship, physicians are less well off than other professionals if they are forced to relocate their practices. For example, attorneys have the benefit of electronic law libraries and e-mail and can temporarily move their work to their home computers. This was the case after Hurricane Katrina, when hundreds of lawyers from New Orleans moved their offices to Baton Rouge and Houston and were able to maintain their clients by using computers and cell phones to maintain contact. For example, the radiologist who has X-ray and mammogram machines cannot easily move this very costly equipment—business interruption insurance is designed to cover the kinds of extra expenses involved in setting up a temporary place of business elsewhere.

Finally, physicians should determine how much income they really can afford to lose if their practice stops. When businesses and factories purchase business interruption insurance, the timeframe for coverage is approximately 1 year because that is how long it might take to relocate a factory temporarily. Because a physician will not need that long to find a temporary office, he or she needs a smaller amount of business interruption coverage.

Business interruption insurance can be as vital to your survival as fire insurance. Most people would never consider opening a practice without buying insurance to cover damage caused by fire and windstorms; however, too many medical practices fail to think about how they would manage if a fire or other disaster damaged their business premises so that they were temporarily unusable. Business interruption insurance is the security that you should have to keep your practice viable and functioning after a disaster.

Review your policies with your agents for exclusions. For example, what would happen to your insurance if a civil authority denied you access to your premises? This happened in New Orleans after Katrina when the mayor ordered evacuation of the city and did not allow the citizens to return for nearly 2 weeks after the storm passed. In addition, one of us (NHB) had a practice on the sixth floor of an office building where the elevator did not work, and there was no potable water for several months, making it impossible to practice in the building. Only a good business interruption policy can protect you and your practice under such circumstances.

EXTRA EXPENSE INSURANCE

Extra expense insurance reimburses your company for a reasonable sum of money that it spends, over and above normal operating expenses, in order to avoid having to shut down during a period of restoration. For example, if your building roof collapses after a heavy thunderstorm, the business interruption policy pays for the income you lost while you could not occupy the building, whereas the extra expense policy covers the rent for a temporary office space while your building is being repaired.

When your practice has to be moved, you will undoubtedly incur many extra expenses, such as rent and installation of telephones. Usually, extra expenses will be paid if they help to decrease business interruption costs. Extra expense insurance is an extremely important add-on to property coverage and is part of most business owners' property insurance. Business interruption insurance coverage reimburses your practice for revenues you lost during down time caused by damage to or loss of your property. Extra expense insurance reimburses you for expenses incurred that are necessary to get your practice back in operation.

BOTTOM LINE

Every doctor has homeowners' insurance, automobile insurance, some form of life insurance, and often disability insurance; however, many of us do not have insurance that will protect us if a disaster occurs. Protect your valuable practice with the insurance that will ensure the viability of your practice if you are struck with a natural or man-made disaster. Although most physicians may conceptually acknowledge the value of disaster planning, many are not ready to make the financial investment to put it in place. For those that are straddling the line between acceptance and spending, keep in mind that when disasters—large or small—strike, the return on investment could literally be the continued operation of your practice. Remember this: "If it ain't broke, don't fix it" may apply to your car but not to your medical practice, which you may have spent your entire lifetime building. Protect it by making sure that it is properly insured.

ACKNOWLEDGMENT

The authors thank Marjorie A. Satinsky, MBA, President of Satinsky Consulting, LLC, a consulting firm in Durham, NC, for her assistance in creating and contributing to this chapter.

Chapter 9

Finding an Alternative Site for Your Practice

Wise men never sit and wail their loss, but cheerily seek how to redress their harms.

—William Shakespeare

Dr. Robert Sharp in New Orleans lost his building after Hurricane Katrina made it uninhabitable. He contacted his hospital, which was able to move his practice to a mobile medical trailer that was set up in the parking lot of the hospital. The trailer had four treatment rooms and small reception area. He shared the trailer with another physician and spent 6 months there until his office building was repaired.

    This doctor was fortunate that the hospital was eager to help him get back in practice. Hundreds of physicians moved their practices into colleagues' offices for several months until they could find new office space or until their building was repaired. Dozens of physicians in the community, however, were not so lucky and had to move and obtain locum tenens employment until their buildings were repaired. There were even others who had to close their practices and seek employment elsewhere. To avoid these problems, take time to answer these questions.

Key Questions

1. Do you have a functional alternate site if a disaster occurs?
2. If your building is uninhabitable, where will you practice?
3. Do you have the equipment that you will need to begin practicing in a new environment?

Some disasters, such as fires, hurricanes, and floods, require a practice to leave the building where you work and move to a new location. This is a daunting task that requires significant planning and execution; however, if you anticipate this possibility, the emotional trauma can be minimized, and the transition can be made to another physical plant in a nearly painless fashion. This chapter discusses the planning that is necessary to find an alternate site for your practice if your building becomes uninhabitable.

Over 40% of businesses (and medical practices) that experience a disaster never reopen, and over a third of those continue to suffer economic difficulties.[1] From major catastrophes to minor crises, unforeseen emergencies can cost your practice hundreds or thousands of dollars in lost revenues. Even a minor disaster can close your practice for days. Whether through a terrorist attack or tornado, disasters and business disruptions happen when you least expect them. Remember that even minor interruptions, such as a power outage, can wreak havoc on your practice. Practices that properly prepare for these threats will rebound faster and suffer fewer financial losses. The best practices with the most foresight are prepared for the worst disasters and thus are able to handle the minor setbacks. Your emergency plans should also identify an alternative practice site where your practice can continue with the day-to-day operations. Rebuilding and maintaining your own fully functioning alternate work site can be very costly; therefore, commercial workplace recovery centers are a good alternative. Paying a monthly fee entitles your practice to declare a disaster and use backup office space that can be tailored to your practice's specific needs, including providing the correct office equipment, exam tables, X-rays, basic laboratory equipment, live phones connected to your telecommunications system, and computer terminals with access to your network and current data.

Before deciding on a workplace recovery provider, make certain that it will allow you to define what constitutes an emergency. Vendors should not require you to meet their definition of an emergency, which might be a much higher threshold than yours.

Although half of businesses estimate that disaster insurance will take less than 3 months for reimbursement, insurance experts and officials estimate at least twice that length of time (see Katrina-staying afloat: http://disasterrecovery.agilityrecovery.com/readysuite/). Now some options will ensure that your practice can be fully operational within 48 hours of an emergency declaration. Whether a disaster is natural or man-made, the effects continue long after. What you do after a disaster is just as important as what you do beforehand. How will you keep your business operational in the face of power outages, flood water, extensive property damage, or other dilemmas?

In some instances, an alternate site is necessary to continue to practice and is to be used when the primary facilities are inaccessible. These alternate sites may be at a different location or address than the normal practice address. (Appendix 11 is a recovery location form that should be prepared before a disaster occurs and can provide a plan if your office, building, or practice becomes unusable.)

CALCULATE THE COST OF DOWN TIME

Although nearly every medical practice will insure against fire, theft, and injury, it is difficult to earmark money for business continuity and disaster recovery contingencies. Perhaps this occurs because of how we have traditionally defined a disaster—as something cataclysmic and highly improbably, like a hurricane, an earthquake, a terrorist attack, or a blackout. If we consider a business disaster as anything that disrupts your ability to see patients and function as a medical practice, then even a truck hitting the telephone pole that supplies electricity and phone to your building now becomes a disaster. Disasters do not have to be horrific storms or fires but could instead be a power outage during a storm; spam or ad ware that brings down your computer systems can also wreck havoc on your practice. A disaster is any occasion when your practice is unable to access and use the building, the data, and systems it needs to function as a medical practice.

To understand the importance for investing in disaster recovery, you first need to understand the true costs of down time—in terms of employees and productivity, equipment, and data. A few calculations can help you to make an estimate of how much down time could cost your practice. Your costs need to be calculated on an hourly, daily, weekly, and monthly basis. With these calculations and an estimation of how long you might not be able to practice, you can determine insurance needs and the kind of alternative site planning you will need. For example, if your overhead costs $1,000,000 per year and your practice is open for 2,000 hours annually, then your cost per hour is $500. If your practice was disrupted by a disaster for 6 weeks, then you would need $120,000 just to maintain your overhead expenses during the rebuilding efforts or the transition to another facility. It is nearly impossible to estimate accurately the true costs and consequences of total business interruption. You need to know that the down time will typically cost your practice much more than disaster recovery or just rebuilding the physical plant. There really is no way to measure the impact and costs of missed practice opportunities, employee dissatisfaction, employee turnover, retraining new employees, damaged reputation, and long-term loss of patients who have gone elsewhere for their medical care.

You can purchase programs that will ensure that your practice can be physically moved to a new facility and may even use a mobile office that can be delivered to your site so that you can practice in the same geographic area. These programs can make use of satellite phone technology and the Internet to make your communications with your patients possible in just a few hours. If a power shortage occurs, these programs can provide power generators to run your electrical needs. If your computers are damaged, they can ship computer hardware, software, and technology directly to your site within hours.

One of these programs, Agility (www.agility.com), provides computers, servers, desks for your office staff, chairs for your reception area, exam tables, Internet access, printers, backup tape drive capability, fax machines, and telephone capabilities all within hours of a disaster. These programs are available for $250 per month. Of course, the costs vary depending on the requirements that you will need to transition your practice.

There is more to business continuity than simply recovering your data and applications after a major or minor business interruption. For example, your employees need to transition quickly their daily routine of answering patient inquiries, scheduling office visits, and refilling prescriptions; however, relocating and acclimating your practice with a new workplace can be overwhelming.

A relatively new alternative to a standard workplace recovery facility involves the use of a mobile worksite. Mobile recovery services bring the entire recovery effort to your door. Using a mobile recovery service can serve as a stand-alone disaster recovery option or can supplement your existing work place recovery plan.

A mobile recovery service provides the technology, space, infrastructure, and technical expertise to put all of your employees back to work after a disaster or disruption. These programs offer fully functional mobile facilities that can move your practice from the disaster zone to the new facility within hours.

These programs are robust and can segue your practice with just a few employees to large multispecialty practices with more than hundreds of employees. They can create a work site that is fully wired for transmission of Internet data and telecommunication. The new site is equipped with the latest model of desktop computers, professional business furniture, heat and air conditioning, and an employee lounge, restroom, and kitchen.

If you want to be fully prepared for a disaster, then you can create a business impact analysis. This can be accomplished by performing a cash flow analysis, whereby the practice estimates its weekly revenue and expenses. Revenue projections can be determined by the percentage of gross charges collected and applied to the total amount of accounts receivable in insurance accounts, as it is unlikely that the practice will receive many payments from patient balance accounts during a sustained period after a major disaster. Assuming the practice does not have business interruption insurance, the practice will have to assess its ability to finance the practice for a period of time through its accounts receivables, business loan, and/or line of credit. A business impact analysis determines in how many days or weeks without your regular income stream you will be able to fund your business operations. This will help to determine how long it will take before the loss of income affects the delivery of your services to your patients. You must also consider how many payroll periods can you meet with no income before you will have to lay off your existing employees.

Next you will need to know what your recovery time objective is, which is that point when a practice expects to be back in operation. For most medical practices, the recovery time objective has to be within hours or days.[2] Patients who are sick and require your services will not be able to wait for long periods until your practice can be operational. In order to determine your recovery time objective, examine each discrete, definable component of your practice—each department and its critical services that you want to resuscitate.

Your recovery time objective will determine what resources you need to purchase or implement. If a quick recovery time objective is dictated, then resources must be spent. For example, if your staff members need electricity to power their computers and lights, then you will have to purchase a generator; thus, you must allocate resources for this. A quick recovery time objective will cost more than a slower recovery time objective.

In many cases, when determining an alternate site for your practice, you may be able to form a mutual partnership with another practice, arranging for an emergency sharing of one another's facilities in case of disaster.

Partner, if possible, with someone who is geographically far enough away to minimize the chances that you will both be affected by the same emergency situation or subsequent events. Allow for the fact that if you have to relocate temporarily to a shared facility, the patients and business operations of the physician(s) who owns the accessible location will take precedence. The efficiency and workload of the displaced practice may be reduced.[3]

Your plan should also detail how you can communicate with vendors about where to deliver supplies, how to notify laboratories about your relocation, and how to inform other service people about how you are coping with your recovery. Also, check with your vendors to see whether they possess adequate business recovery plans. If they do not, urge them to put one together or find another vendor who is adequately prepared.

ALTERNATE SITE DEVELOPMENT PLAN

Begin by prioritizing your practice's operational needs. Determine your practice's critical needs in the areas of human resources, technology, telecommunications, physical environment, operational resources, and infrastructure support. This will help you to prioritize specific people, systems, and facilities that must be brought back online first. It will also help you to assign the budget and resources necessary to safeguard these assets and those critical employees who are required to take action quickly.

Next, understand the risks to your practice. Assess your risks and face your areas of exposure head on. Remember that it is not just hurricanes and floods that create disasters. Sprinkler malfunction, theft, equipment damage, and power outages are also plausible risks to nearly every practice, regardless of their geography. In addition to geography, each business has specific vulnerabilities that need to be addressed; consider security, inventory, Internet availability, and data storage.

Develop and document a realistic plan, as discussed in Chapters 2 and 3. By documenting a practice continuity strategy, you make the leap from well-intentioned, employee lounge talk to an actionable, funded plan. Because physicians and their practices have specific operational requirements and therefore different needs when it comes to disaster recovery, your plan must be tailored to your unique situation. Do not settle for generic, fill-in-the-blank plans. Take the time to customize and document a plan that will work specifically for your practice.

The best plans for an alternate site location include selecting your vendor partners in advance of the disaster. In the face of a crisis, you will not want to worry about evaluating and negotiating with vendors that will participate in your practice's recovery process. By pre-establishing contractual relationships with reputable vendor partners, you increase your ability to identify, respond, and resolve potentially disastrous situations.

Also, be sure to establish a protocol for crisis situations. As we witnessed in New Orleans, documents alone do not equate into action. You need commitments from your partners and staff to initiate the action plan. An internal protocol must be established, and a chain of command must be assigned that clearly gives authority to the key leaders who can declare the disaster and set the wheels of your disaster recovery plan in motion.

Finally, practice, prepare, and run drills so that everyone is familiar and comfortable with the plan. Unfortunately, disasters and crises often lead to stress, blame, and breakdowns, as we continue to see in the wake of Katrina both locally in New Orleans and also at the state and federal levels. You, your staff, and your vendor partners cannot rise to the challenge if any of you are reading the disaster plan for the first time *after* disaster strikes. For any team to work well together, preparation, coordination, and trust are required. Grace under pressure will occur only if your entire team has had a chance to go through the motions and put theory into practice *before* the crisis sets in.[4]

BOTTOM LINE

Predicting a terrorist attack, a natural disaster, or a power outage is impossible. You can do only so much to prepare your practice for catastrophe, but that just might be enough to secure your bottom line. Part of every disaster plan includes having an alternate practice site in case your facility is no longer functional. This can be easily accomplished with another group, or you can contract with a mobile recovery service that will have your practice running in 48 hours. In the next chapter, we will discuss how you can protect your employees by encouraging them to have their own disaster plan.

REFERENCES

1. Economist Intelligence Unit. "Catastrophe Risk Management: Preparing for Potential Storms Ahead," an economist intelligence unit white paper sponsored by ACE, IBM, and KPMG. Available at: http://www.kpmg.com/nr/rdonlyres/240500fe-fb0c-4ec2-a50f-b3547cb44d3e/0/catastropheriskmanagement.pdf, accessed February 11, 2008.
2. Nonprofit Coordinating Committee of New York. "Disaster Planning, Emergency Preparedness and Business Continuity." Available at: www.npccny.org/infor/Disaster_Planning.doc, accessed February 11, 2008.
3. Cascardo, D. C. "Practice Prescriptions—Preparing Your Medical Practice for Disaster from Medscape Money and Medicine." Available at: www.nymgma.com/files/NewsletterFall_04.pdf, accessed February 11, 2008.
4. Alcorn, R. J. Chief Operating Officer, nFrame. "An Ounce of Disaster Preparedness Is Worth a Pound of Business Continuity." Available at: http://www.nframe.com/PDF/OunceOfDisasterPreparednessFinal.pdf, accessed February 11, 2008.

Chapter 10

Disaster Planning for the Employees

When you live prepared, you're prepared to live!

—Matt Lawrence, What to Do Til the Cavalry Comes.
iUniverse, 2006.

In addition to having a disaster plan available for our practices, our employees must be encouraged to have their own personal disaster plan. If your employees are not ready to return to work, you will not be able to implement your practice's plan; therefore, you must make every effort to motivate your employees to have a disaster plan that takes care of their home and their families. This chapter covers the disaster planning for your staff, which might also include your office manager and physicians, who also need to have a personal disaster plan. It also provides suggestions for staff members when creating their own disaster plan.

Chapter 10: Disaster Planning for the Employees

KEY QUESTIONS

1. Do your employees have a personal disaster plan for themselves and their families?

2. Do your employees know where to go if a disaster impacts the building, the hospital, or your entire community?

3. Do your employees have several days of supplies so that they can manage their family if a disaster makes it impossible to obtain food and water?

4. Do your employees have a plan if their house is destroyed and if they have to seek shelter elsewhere?

If the answer to any of these questions is no, then you might want to distribute a copy of this chapter to your employees and then make sure that they have a personal disaster plan in place.

For most large natural disasters, your employees will need to plan to be on their own for a minimum of 72 hours.[1] This is usually the response time for national, state, and local governments to get organized and coordinated before beginning to restore order in the community. Because of Katrina in 2005, we see that it was approximately 72 hours before the National Guard was called in to help restore law and order and to begin helping the citizens, doctors, and hospitals to begin functioning again.

First the employees must make their homes safe. Emergency response services will likely not be able to respond immediately to the needs of your employees; thus, they must be prepared to take care of themselves and their families for at least a few days until help arrives.

Each employee should have a plan for their household. The form that is shown in Figure 10-1 should be completed by someone in the household and updated at least annually.

An out-of-area contact person needs to be designated. He or she should live far enough away to not be affected by the same disaster as the employee and must have the names and contact information (i.e., cell phone number and e-mail address) of the employee so that he or she can inform others of the employee's situation in the disaster

Key Questions | 155

Family Emergency Plan

Prepare. Plan. Stay Informed.

Make sure your family has a plan in case of an emergency. Before an emergency happens, sit down together and decide how you will get in contact with each other, where you will go and what you will do in an emergency. Keep a copy of this plan in your emergency supply kit or another safe place where you can access it in the event of a disaster.

Out-of-Town Contact Name: _____ Telephone Number: _____
Email: _____
Neighborhood Meeting Place: _____ Telephone Number: _____
Regional Meeting Place: _____ Telephone Number: _____
Evacuation Location: _____ Telephone Number: _____

Fill out the following information for each family member and keep it up to date.

Name: _____ Social Security Number: _____
Date of Birth: _____ Important Medical Information: _____

Name: _____ Social Security Number: _____
Date of Birth: _____ Important Medical Information: _____

Name: _____ Social Security Number: _____
Date of Birth: _____ Important Medical Information: _____

Name: _____ Social Security Number: _____
Date of Birth: _____ Important Medical Information: _____

Name: _____ Social Security Number: _____
Date of Birth: _____ Important Medical Information: _____

Name: _____ Social Security Number: _____
Date of Birth: _____ Important Medical Information: _____

Write down where your family spends the most time: work, school and other places you frequent. Schools, daycare providers, workplaces and apartment buildings should all have site-specific emergency plans that you and your family need to know about.

Work Location One
Address: _____
Phone Number: _____
Evacuation Location: _____

School Location One
Address: _____
Phone Number: _____
Evacuation Location: _____

Work Location Two
Address: _____
Phone Number: _____
Evacuation Location: _____

School Location Two
Address: _____
Phone Number: _____
Evacuation Location: _____

Work Location Three
Address: _____
Phone Number: _____
Evacuation Location: _____

School Location Three
Address: _____
Phone Number: _____
Evacuation Location: _____

Other place you frequent
Address: _____
Phone Number: _____
Evacuation Location: _____

Other place you frequent
Address: _____
Phone Number: _____
Evacuation Location: _____

Important Information	Name	Telephone Number	Policy Number
Doctor(s):			
Other:			
Pharmacist:			
Medical Insurance:			
Homeowners/Rental Insurance:			
Veterinarian/Kennel (for pets):			

Dial 911 for Emergencies

Figure 10–1 Family emergency plan.

continues

Chapter 10: Disaster Planning for the Employees

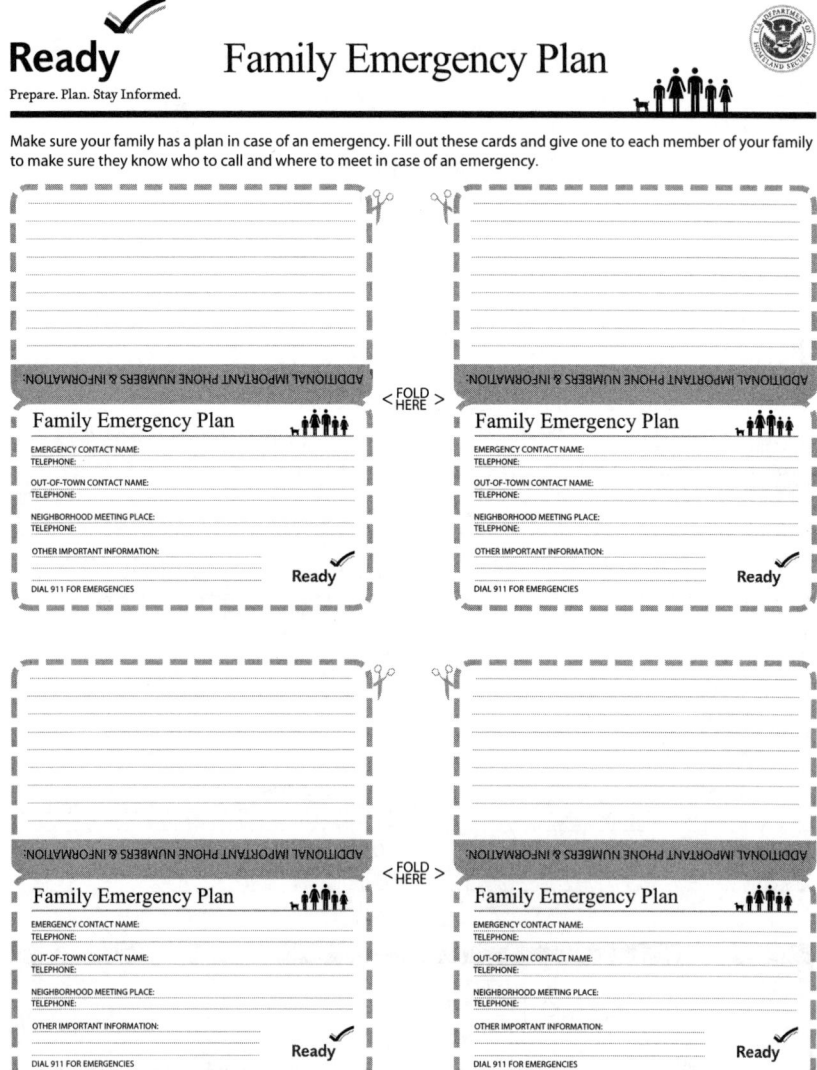

Figure 10–1 Family emergency plan (*continued*).
Source: Courtesy of Ready.gov.

area. The out-of-area person needs to be available for others to call; then the employee can pass status information on to other family members and friends, letting them know that he or she is safe, that he or she needs assistance, and his or her location.

Each employee should duplicate important documents and have a written inventory, photographs, and a video of their possessions. This video should not be kept in the home but instead in a safety deposit box or with someone that is trustworthy. It should be located at a distance from the employee's home. Important documents might include passports, driver licenses, social security card, wills, deeds, recent W-2 forms, financial statements, insurance information, marriage licenses, and a list of prescription medications that family members use. Doctors need a copy of their state medical license, proof of malpractice, and a copy of their medical school diploma.

Employees should be advised to make their homes safe. They should think and look with a "disaster eye" and identify and fix any possible hazards. For example, smoke detectors need to be installed, and the batteries should be changed every 6 months. The hallways and doors need to be cleared so that members of the household can easily exit the home in an emergency. A fire extinguisher should be kept on each floor, and each member of the family should know how and when to use it. All flammable or highly reactive chemicals, such as bleach, ammonia, and paint thinners, should be securely closed and separated from each other. Use restraints such as bungee cords to secure heavy items like bookcases and file cabinets. Know where the gas main and other utilities are located and how and when to turn them off. If the home has window safety bars, emergency releases need to be intact. Finally, make certain that the house's street number is clearly visible from the street so that emergency vehicles can find you.

It is also important to make a household/family plan. All of the employees need to talk to other family members about potential disasters and why it is necessary to prepare. Make every effort to involve each member of the family in the planning process.

Everyone in the family needs to know who the out-of-area contact is. If a family member cannot reach the home or is out of the area, then he or she has a central contact person who can help coordinate the location and safety of the family. Every family member needs to know the evacuation route in case the head of the household is not at home and cannot be reached. Each family member should carry an emergency response card that contains this vital information. Be sure that you remember the special needs of children, seniors, and those with disabilities. For those family members who need medication, a list of their medication should be on their emergency response cards.

No personal disaster plan is complete without a disaster supply kit or travel bag for each family member in case evacuation is necessary. Experts suggest that you should be prepared to be self-sufficient for at least 3 days. Store your household disaster kit in an easily accessible location. Put the contents in a large, watertight container such as a large plastic garbage can with a lid and wheels. The container should be easy to move and load if you need to leave.

BUILD A DISASTER KIT

After a major disaster, the usual services that we take for granted (such as running water, refrigeration, and telephones) may be unavailable. The following lists of items will help you put together your emergency kit. Bring these lists with you to the grocery and hardware stores to supplement any of these items that you do not already have at home.

- Water—1 gallon of drinking water per person per day (see the section about water)
- Food—ready to eat or requiring minimal water (see the section about food)
- Manual can opener and other cooking supplies
- Plates, utensils, and other feeding supplies
- First-aid kit and instructions
- A copy of important documents and phone numbers
- Warm clothes and rain gear for each family member

- Heavy work gloves
- Disposable camera
- Unscented liquid household bleach and an eyedropper for water purification (see the section on water)
- Personal hygiene items, including toilet paper, feminine supplies, hand sanitizer, and soap
- Plastic sheeting, duct tape, and utility knife for covering broken windows
- Tools such as a crowbar hammer and nails, staple gun, adjustable wrench, and bungee cords
- Blanket or sleeping bag
- Large heavy duty plastic bags and a plastic bucket for waste and sanitation
- Diapers and other items for babies and small children
- Special-needs items for family members with mobility issues, such as an extra cane or manual wheelchair in case there is no power for recharging and an electric wheelchair

A component of your disaster kit is your travel bag. Put the following items in a backpack or another easy-to-carry container in case you must evacuate quickly. Prepare one travel bag for each family member, and make sure that each has an identification tag. You may not be at home when an emergency strikes; therefore, keep some additional supplies in your car and at work, taking into consideration what you would need for your immediate safety.

- Flashlight
- Radio—battery operated
- Batteries
- Whistle
- Dust mask
- Pocket knife
- Emergency cash in small denominations and quarters for phone calls
- Sturdy shoes, a change of clothes, and a warm hat
- Local map

- Some water and food
- Permanent marker, paper, and tape
- Photos of family members and pets for identification purposes
- List of emergency point-of-contact phone numbers
- List of allergies to any drug (especially antibiotics) or food
- A copy of health insurance and identification cards
- Extra prescriptions for eye glasses, hearing aid, or other vital personal items
- Prescription medications and first-aid supplies
- Toothbrush and toothpaste
- Extra keys to your house and vehicle
- Your child's favorite toy, game, or book, as well as his or her emergency card with reunification location and out-of-area contact information

For information on how to prepare a travel bag for your pets, see the section about extra tips for pet owners.

For those households that have seniors and household members with disabilities, additional planning steps are necessary. Seniors or those with special needs should have someone assigned to check on them and be responsible for their safety. If you have family members that are receiving assistance from a home health care agency or in-home support provider, find out how the provider will respond in an emergency. You need to have a backup or alternative provider who you can contact in an emergency if the primary agency or provider is unable to function or provide assistance. If you have family members who require a wheelchair, you need to have a plan on moving the wheelchair as well as the disabled person. If the person requires a large or heavy motorized wheelchair, you might need to have a manual wheelchair as a backup. For those who are hearing impaired, it is a good idea to have extra batteries for the hearing aids with the emergency supply kit.

The employee's children need to know basic personal information in case they become separated from a parent or guardian. An emergency card with information for each child, including his or her full name, address, phone number, parent's work number, and out-of-area contact information, should be prepared for each child.

Know the policies of the school or daycare center that your children attend. Make plans to have someone pick up your children if you are unable to do so and make sure the school or center has a list of those authorized to do so. Keep your child's school updated with current emergency contact information. Each child should know the family's alternate meeting site, as well as the out-of-area contact person. Children should know how to dial 911.

PETS

For those employees who have pets, be aware that most disaster centers cannot accept pets because of health and safety regulations. The only exceptions are service animals for people with disabilities. Some large communities have set up temporary shelters for animals in close proximity to human shelters. Employees should arrange for a neighbor to check on your pets and take care of them if a disaster occurs while you are not at home. If necessary, plan for a friend or relative outside of the disaster area to shelter your pet. Keep your pet's identification tags up to date and clearly visible on the pet's collar. You might even consider having your pet microchipped. Consider making a disaster kit for your pet, including leashes, carriers to transport the pet, photos of the pet, food and water for at least 3 days, bowls, pet toys, a list of the pet's medical condition, pet medications, feeding schedule, and your veterinarian's phone number in case you have to board your pet.

FOOD

Ample food and water will be needed after a disaster until help arrives. Food items that are stored should be according to your family's desires and tastes. Another alternative is to buy meals ready to eat, or MREs. These prepared meals are available at www.MREfoods.com. Do not forget to take into consideration any dietary restrictions and preferences you may have. The ideal foods for a disaster kit include shelf-stable foods that do not require refrigeration. Foods should not require cooking and minimal preparation. Food in your disaster kit should be rotated by date to avoid having to eat expired food. Most canned food can be stored for at least 18 months. Low-acid foods like meat products, fruits, or vegetables will normally last at least 2 years. Dried products such as boxed cereal, crackers, cookies, and dried milk will last for 6 months. The food should be stored in airtight, pest-resistant containers in a cool, dark place. If you experience a power outage, you can use the food in the refrigerator if you keep the door closed and do not open it very often. Food in the freezer will normally remain safe for 2 days.

WATER

In a disaster, you may not have access to water, as it may be cut off or contaminated; therefore, you need store enough water for everyone in your family for 3 days. The recommendation is to store 1 gallon of water per person, per day. This amount is adequate for general drinking purposes. Do not forget to provide adequate water for your pets. If you store tap water, use the food-grade plastic containers that are available in most sporting goods stores. Replace the water in these bottles every 6 months. If you are storing commercially bottled drinking water, replace it at least once a year.

If you run out of stored drinking water, you will need to treat the water before using it. Begin by straining any large particles of dirt by pouring the water through a couple of layers of paper towels or clean cloth. Next, purify the water one of two ways:

1. Boil it. Maintain a rolling boil for 3 to 5 minutes. To improve the taste, pour it back and forth between two clean containers to return the oxygen to the water.
2. Disinfect the water. If the water is clear, add 8 drops of bleach per gallon of water.
    If it is cloudy, then add 16 drops of bleach per gallon. Then shake or stir the water and allow it to stand for 30 minutes. A slight chlorine taste and smell is normal after adding bleach.

UTILITIES

Gas

Natural gas can leak from your system and cause an explosion inside your home. If you smell gas or see a broken gas line, then immediately shut off the main valve, and open all windows and doors. Never use candles or matches if you suspect a gas leak. Instead, find the main shutoff valve, which is usually located on the gas line coming into the main gas meter outside of the house. Most valves work by turning the valve in either direction until the lever is perpendicular to the direction of the pipe (see Figure 10-2).

To turn the valve, keep a crescent wrench or gas shutoff tool nearby to turn the lever. After you turn off the gas, never attempt to turn it back on yourself. Wait for your utility company to do it, but be aware that it may take several days for it to be turned back on.

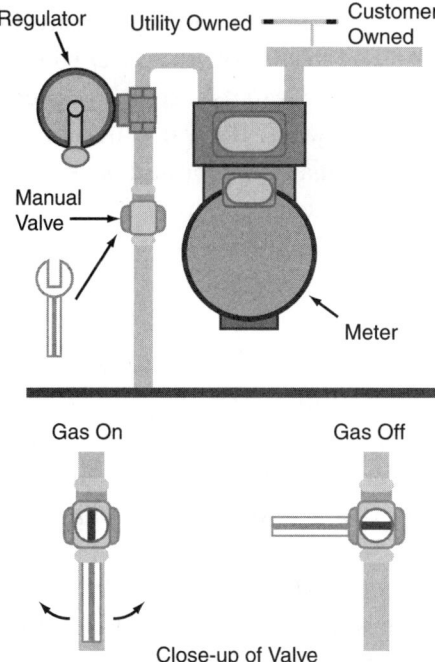

Figure 10-2 Natural gas safety and shut off valves.

Electricity

Know the location of your home's main electric switch, which is usually in the garage or outside of the home where the power lines enter the household. The panel box may have a flip switch or pull handle on a large circuit breaker. The electricity should be shut off when you see arcing or burning in the electrical device or any electrical appliance, if there is a fire or significant water leak, if you smell burning insulation, if the area around plugs or switches is blackened or hot to the touch, or if a complete power loss is accompanied by the smell of burning material.

If there is a power outage, turn off and unplug all appliances and computers. Then leave one light on to indicate when power has been restored. Do not use a gas stove for heating or operate generators indoors or in the garage, as this could result in carbon monoxide poisoning.

Water

Water leaks can cause property damage and create an electrocution hazard; therefore, be sure to shut off the water when there is a water leak inside the building. The water shutoff valve is usually located in the basement or garage or where the water line enters the home. The water shutoff is located on a riser pipe and is usually a red or yellow wheel. Turn the wheel clockwise to shut off the water to the home.

Phone

No disaster plan is complete unless you know how you will be communicating with other family members, your office, and the person that is responsible for the disaster recovery at the medical practice. Avoid making nonurgent phone calls after a disaster, as there is always increased phone traffic that occurs, possibly leading to jammed phone circuits. Also, do not depend on your cell phone, as increased cell phone traffic can quickly overload wireless capacity; therefore, record an outgoing message on your voicemail so that callers can be reassured of your safety. A good suggestion is to keep coins in your disaster kit, as pay phones are more likely to work before other phone lines. If you have a cordless phone or phone system, electricity is required for them to function; thus, you will need a backup phone system that does not require electricity.

EVACUATION

If you smell gas or smoke, see fire, or fear for your safety, then immediately evacuate all occupants in the house. Lock your home, and shut off the water and electricity. From a safe location, call 911 to report the incident. If local officials issue evacuation orders, use the evacuation routes and methods that the officials recommend.

FIRE

If your smoke detector goes off or if you see a fire, remain calm and leave the home.

If you see smoke under the door, find another way out. In order to determine the situation on the other side of a closed door, feel the door with the back of your hand before you open it. If it is hot, find another way out. If there is smoke or fumes in the room, drop to the floor to avoid smoke and fumes, and then crawl to safety. If your clothes catch on fire, stop where you are, drop to the ground, and roll back and forth to smother the flames.

After you are safely out of the house, call 911 from a safe location. If you are trapped in a burning building, stay near a window and close to the floor. If possible, signal for help.

DISASTER SHELTERS

Immediately after a large disaster, suitable shelter sites will be selected from a predetermined list based on areas of need and estimated numbers of displaced persons. Each site must be inspected for safety before being opened to the public.

Some communities will announce in advance where the disaster shelters are located; others will declare the location after the disaster has occurred. As soon as disaster sites have been formally designated, this list will be announced through local media to the public. If it is unsafe to shelter-in-place and you do not have an alternative, evacuate to a designated emergency shelter.

If you do go to a shelter, be sure to tell your out-of-area-contact where you are going. When you go, take your travel bag with you to the shelter. Initially, emergency shelters may not be able to provide basic supplies and materials. Consider bringing extra items (e.g., blanket, pillow, air mattress, towel, washcloth, diapers, food, and supplies for infants). Make separate plans for your pets. Only service animals are allowed in "human" shelters.

BOTTOM LINE

In order for an employee to be effective at the work place, he or she has to be effective in the home. Each employee must be able to protect his or her family first before he or she can focus on the medical practice. We have laid out a plan that almost every employee can live by. Put your personal disaster plan in place, and then you can help get your practice back functioning again.

Reference

1. "How Would You Survive for 72 Hours?" Available at: http://www.72hours.org, accessed February 11, 2008.

Chapter 11

Conclusion

To know and not to do is not to know!
Knowledge does not become power until it is used.

—Harvey Mackay, author of
Swim With the Sharks Without Being Eaten Alive

Hopefully those who read this book will not have to make use of their newly created disaster plans. Unfortunately, many physicians and practices will, as disasters do occur and can wreck havoc on any practice; however, with preparation, planning, and careful execution, the impact can be minimized. We are certainly aware that geography and climate cannot be changed, but vulnerability to a disaster can be significantly reduced by following the suggestions in this book.

Unfortunately, many practices learn about the advantages of disaster planning the hard way (i.e., *after* the disaster has occurred). Hopefully this book demonstrates that a disaster does not have to become a full-fledged catastrophe with loss of employees, complete interruption of patient care, and loss of income. The impact of a disaster can often be mitigated or avoided altogether by a comprehensive, systematic emergency-preparedness program. The plan that we have provided creates a means for recognizing and preventing risks and for responding effectively to emergencies.

An increasing number of physicians know that small-scale emergencies can be contained if doctors and staff are prepared to react quickly. Damage can be limited even in the face of a large-scale disaster.

Disaster planning should not take place in a vacuum. To work effectively, it must be integrated into the practice's routine operating procedures. In fact, when planning for disasters, you will also be working toward the accomplishment of other goals, such as better communication between the staff members and with patients.[1]

Do not forget the three important characteristics of an effective disaster plan: comprehensiveness, simplicity, and flexibility. The plan needs to address most types of emergencies and disasters that your practice is likely to face, including plans for both immediate response and long-term recovery efforts.[1]

This book has shown you how to create a plan that is easy to follow. Doctors and staff who are faced with a disaster often have trouble thinking clearly and will be out of their comfort zone after a disaster. Few will have had previous experience dealing with a disaster and the recovery efforts; therefore, concise instructions and training are critical to the success of the plan. The key is to create a clear, simple plan without sacrificing comprehensiveness. Above all, remember that you cannot anticipate every detail; thus, be sure that although your plan provides basic instructions it also allows for some on-the-spot creativity.

No matter how much effort you have put into creating the perfect disaster plan, it will be largely ineffective if your staff is not aware of it, if it is outdated, or if you cannot find it during a disaster. A concentrated effort must be made to educate and train staff members in emergency procedures. Each person should be aware of his or her responsibilities, and regular drills should be conducted if possible. Keep several copies of the plan in various locations, including offsite (ideally in waterproof containers). Each copy of the plan should indicate where other copies may be found.

Most important, the disaster plan must be updated periodically. If a plan is outdated, it may not be able to help you deal effectively with disasters.

Disaster planning is essential for any medical practice to provide the best possible care for your patients and your staff. Disaster can strike at any time—on a small or a large scale—but if a practice is prepared, the damage may be decreased or avoided. A disaster plan must be considered a living document. Its risk-assessment checklist must be periodically reviewed. Its lists of staff and equipment must be updated and its priorities revised as needed. An effective disaster plan will ensure that patients receive continuity of care, that the practice experiences the minimum of down time, and that the loss of income is also kept to a minimum.

The probability of a disaster impacting your practice is highly uncertain. A disaster plan, however, is similar to liability insurance: It provides a certain level of comfort in knowing that if a major catastrophe occurs, it will not result in financial disaster for your practice. Insurance alone is not adequate because it may not compensate for the incalculable loss of business during the interruption or the patients that never return. The consequences of a disaster are unpredictable and often unimaginable. Certainly, no disaster plan will cover all disaster situations; however, the plan that we have suggested will work for the most common disasters, both natural and manmade. Of course, the best advice is to have a plan in place long before a disaster occurs. Then update it on a regular basis, and make the necessary modifications so that the plan fits the needs of your practice.

Now you know what to do. We have given you a plan that you can create and then implement if a disaster occurs. Now, please, "Just do it!"

REFERENCE

1. Patkus, B. L., Motylewski, K. "Emergency Management, Disaster Planning." Northeast Document Conservation Center. Available at: http://www.nedcc.org/resources/leaflets/3Emergency_Management/03 DisasterPlanning.php, accessed February 11, 2008.

Appendices

Appendix 1: **Employee Contact List** [See also Figure 3-3]* 173

Appendix 2: **Key Contacts** [See also Figure 3-4]* 175

Appendix 3: **Critical Business Functions**
[See also Figure 3-5]* 177

Appendix 4: **Computer Equipment and Software Form**
[See also Figure 3-6]* 179

Appendix 5: **Voice/Data Communications Form**
[See also Figure 3-7]* 181

Appendix 6: **Miscellaneous Resource Form**
[See also Figure 3-9]* 183

Appendix 7: **Disaster Response Checklist Form**
[See also Figure 3-10]* 185

Appendix 8: **Emergency Phone Numbers**
[See also Figure 3-11]* 187

Appendix 9: **Vital Records** [See also Figure 3-12]* 189

Appendix 10: **Corporate Headquarters Telephone Numbers** ..191

Appendix 11: **Practice Recovery Work Area Checklist** 193

Appendix 12: **Resources Required Over Time** 195

Appendix 13: **Travel Request Form** 199

Appendix 14: **Recovery Boxes** 201

Appendix 15: **Critical Resources to Be Retrieved** 205

Appendix 16: **Personnel Location Control Form** 209

Appendix 17: **Status Report Form** 211

Appendix 18: **Activity Schedule** 213

Appendix 19: **Guide to Record Retention** 217

*Chapter 3 and appendix forms from that chapter are printed with permission by the Institute for Business & Home Safety as a derivative work of the *Open for Business*® toolkit at www.disastersafety.org.

Appendix 1

Employee Contact List

Name of employee: _____
Position: _____
Key responsibilities: _____
Home address: _____
City, State, Zip: _____
Home phone: _____
Cell phone: _____
Office phone: _____
Pager/beeper: _____
Home e-mail: _____
Work e-mail: _____
Emergency contact: _____
Relationship: _____

Appendix 2

Key Contacts

Name of business or service: _____
Account number: _____
Password: _____
Materials/service provided: _____
Street address: _____
City, State, Zip: _____
Company/service phone: _____
Primary contact: _____ Title: _____
Primary contact phone: _____ Contact cell phone: _____
Primary contact pager: _____ Contact fax: _____
Primary contact e-mail: _____ Contact website: _____
Alternate contact person: _____ Title: _____
Alternate contact phone: _____ Alternate's cell phone: _____
Alternate contact pager: _____ Alternate's fax: _____
Alternate contact e-mail: _____
Recovery notes: _____

Appendix 3

Critical Business Functions

Appendix 3: Critical Business Functions

Practice function: _____

Priority: ☐ High ☐ Medium ☐ Low

Employee or physician in charge: _____

Timeframe or deadline: _____

Business function: _____

Priority: ☐ High ☐ Medium ☐ Low

Employee or physician in charge: _____

Timeframe or deadline: _____

Practice function: _____

Priority: ☐ High ☐ Medium ☐ Low

Employee or physician in charge: _____

Timeframe or deadline: _____

Brief description of procedures to complete function: _____
_____
_____

You should consider writing out two scenarios, one for a short disruption (i.e., several days) and the other for a more lengthy disruption (i.e., weeks or months).

Recovery note: _____
_____
_____

Appendix 4

Computer Equipment and Software Form

Appendix 4: Computer Equipment and Software Form

Name of Vital Record:

Media:	☐ Network	☐ Print version
	☐ Hard drive	☐ Microfilm
	☐ Laptop	☐ Internet
	☐ CD	☐ Other
	☐ Diskette	Explain:

Is it backed up?   ☐ Yes   ☐ No

How often is it backed up?	☐ Hourly	☐ Quarterly
	☐ Daily	☐ Semi-annually
	☐ Weekly	☐ Yearly
	☐ Monthly	☐ Never

Where is it stored?

Can the record be recreated?   ☐ Yes   ☐ No

Has the back up been tested?   ☐ Yes   ☐ No

Date of backup test? _____

Practice function it supports: _____

_____

_____

Recovery notes: _____

_____

_____

Appendix 5

Voice/Data Communications Form

Appendix 5: Voice/Data Communications Form

Type of Service:
- ☐ Telephone
- ☐ PBX w/ ACD*
- ☐ PC data communications
- ☐ Cell phone
- ☐ Fax machine
- ☐ Two-way radio & pager
- ☐ Other
  Explain: _____
  _____

Description and Model Number: _____

Status: ☐ Currently in use   ☐ Will lease/buy for recovery location

Voice Communication Features:
- ☐ Voice mail
- ☐ Speaker
- ☐ Conference
- ☐ Conversation recorder
- ☐ Other
  Explain: _____
  _____

Data Communications Features:
- ☐ Cable
- ☐ DSL
- ☐ T1
- ☐ Dial-up
- ☐ Other
  Explain: _____
  _____

Quantity: _____

Primary supplier/vendor: _____

Alternate supplier/vendor: _____

Recovery install location: _____

Recovery notes: _____

*Automatic Call Distribution

Appendix 6

Miscellaneous Resource Form

Appendix 6: Miscellaneous Resource Form

ITEM	Quantity	Vendor/Supplier	Alternate Vendor/Supplier
Desks			
Chairs (reception)			
Cabinets			
Exam tables			
Chairs (exam room)			
Paper towel dispensers			
Wastebaskets			
Copy machine			
Fax machine			
Telephones			
Modem			
Surge protector			
Power strips			
Disposable gloves			

Appendix 7

Disaster Response Checklist Form

Appendix 7: Disaster Response Checklist Form

Water. If storage space allows, store 2 gallons of water per person per day for drinking and sanitation. Store in plastic containers or use commercially bottled water.

Food and utensils. Have at least a 1- to 3-day supply of nonperishable food, which might include ready-to-eat meats, juices, and high-energy foods such as granola or power bars. Also include a can opener.

NOAA weather alert battery-powered radio and extra batteries.

Fire extinguisher.

AM/FM radio (battery operated with extra batteries).

Flashlight and extra batteries. Do not use candles or open flames during an emergency.

Whistle to signal for help.

Dust or filter masks. Readily available surgery masks will work fine and are available at your hospital.

Moist towelettes for sanitation.

Bleach. Use in the toilet if the toilet is not working.

Basic tool kit, including wrench, hammer, and pliers to turn off utilities.

Broom, shovel, and working gloves.

Plastic sheeting and duct tape to "seal the room."

Medications to include prescription and nonprescription medications such as pain relievers, antacids, and antihistamines.

First-aid supplies, including an assortment of bandages, ointments, gauze pads, cold/hot packs, tweezers, scissors, hemostats, band-aids, gauze, nonadherent sterile pads (various sizes), paper and cloth tape, antibacterial ointment, burn cream, pocketknife (Swiss Army variety), razor blades, large cotton cloth (use for sling, tourniquet, bandage), nonaspirin pain reliever, chemical ice pack, hand warmer packets, safety pins (various sizes), needles, heavy thread, matches (waterproof), eye wash, hand wipes (antiseptic), cotton balls, cotton pads, alcohol swabs, and iodine (bottle or pads).

Blankets.

Battery-operated fans.

Garbage bags and plastic ties for personal sanitation.

Paper supplies, note pads, markers, pens, pencils, plates, napkins, and paper towels.

Disposable camera to record damage.

Cash/ATM and credit cards. Keep enough cash for immediate needs.

Dehumidifier.

Metal cart.

Flashlights.

50-foot extension cord (grounded).

Portable electric fan.

Wet vacuum.

Freezer or wax paper.

Plastic trash bags.

Plastic buckets and trash can.

Paper towels.

Sponges.

Mop.

Monofilament nylon (fishing) line.

Broom.

Gloves (rubber and leather).

Rubber boots and aprons.

Safety glasses.

Plastic sheeting (stored with scissors and tape).

Multi-KV generator.

Safe or locked box.

Copy of employee contact form.

Appendix 8

Emergency Phone Numbers

Appendix 8: Emergency Phone Numbers

Fire department _____

Police department _____

Building supervisor _____

Local ambulance service _____

Hospital (closest) _____

Hospital security _____

Hospital (alternate) _____

Insurance provider/agent _____

- Contact phone _____
- Policy number _____
- Headquarters phone/contact _____

Telephone company _____

Gas/heat company _____

Electric company _____

Water company _____

Red Cross _____

FEMA _____

Radio station(s) _____

_____

_____

Television station(s) _____

_____

_____

Newspaper _____

Appendix 9

Vital Records

- Copy of 3 years of tax returns; 1 year of personal tax returns on principles (affiliates with greater than 20% interest).
- One year of tax returns on affiliated business entity (i.e., ancillary services such as computed tomography scanner, pathology labs, X-ray companies in which the practice is invested).
- For sole proprietorships: a copy of 3 years tax returns with Schedule C.
- List of creditors/contact information with account numbers.
- Sole proprietorships, corporations, and partnerships all need the following:
  - Copy of current profit and loss statement (current within 90 days)
  - Copy of listing of inventory
  - Copy of schedule of liability
  - Copy of balance sheet (as recent as possible)
  - Copy of all of your required licenses (city, occupational, sales tax, federal ID)
  - Copy of doctors' malpractice insurance
  - Copy of doctors' state licenses
  - Copy of doctors' medical school diplomas

Appendix 10

Corporate Headquarters Telephone Numbers

Appendix 10: Corporate Headquarters Telephone Numbers

Name	Position/Title	Office/Cell Phone Number
_____	_____	_____
_____	_____	_____
_____	_____	_____
_____	_____	_____
_____	_____	_____
_____	_____	_____
_____	_____	_____
_____	_____	_____

Reprinted with permission from the Disaster Recovery Journal. Available at: http://www.drj.com/new2dr/recoveryteams-e_pearce.pdf.

Appendix 11

Practice Recovery Work Area Checklist

Appendix 11: Practice Recovery Work Area Checklist

Work Area Scenarios

The Practice Manager will provide the team leader with a work area for the team to use. One of the following is the most likely scenario that will take place.

Work area at the location, if the facility is accessible.

> The Practice manager will provide information about what area the team can use.

Work area at a vendor Practice Recovery Site, if the site is not available.

> The Practice Manager will provide information about what area to use and the estimated time before terminals and communications to the backup site will be available.

Work Area Optimum Requirements

The following lists the minimum requirements for the team at the work area recovery location. Copiers and FAX machines will be available at the work area.

Space in square feet: _____

Office Furniture: Desks: _____   Chairs: _____   File Cabinets: _____

Other Furniture: _____

Telephone Equipment

> Phone Type: _____   Number of Phones: _____

Computer Equipment:

> *Indicate what terminals and PC's would require connection to the network.*

Platform: _____   Terminal Type: _____   Number: _____   Network: _____

PC Software: _____

Reprinted with permission from the Disaster Recovery Journal Available at: http://www.drj.com/new2dr/recoveryteams-e_pearce.pdf.

Appendix **12**

Resources Required Over Time

The following two forms are used to plan the arrival of recovery resources to the Work area. List only the increased amounts in each column. For example, the team needs 35 people over all. They assign 15 at the 24 hours slot, another 5 in the 48 hours slot and 15 more in the 72 hours slot.

Appendix 12: Resources Required Over Time

Resources Required Over Time						
Function / Resources	24 hours	48 hours	72 hours	1 week	2 weeks	1 month
Function Name						
Staff						
Area size						
Desks						
Chairs						
Telephones						
Faxes						
PCs						
Printers						
(Other)						
Function Name						
Staff						
Area size						
Desks						
Chairs						
Telephones						

Function / Resources	24 hours	48 hours	72 hours	1 week	2 weeks	1 month
Faxes						
PCs						
Printers						
(Other)						
Function Name						
Staff						
Area size						
Desks						
Chairs						
Telephones						
Faxes						
PCs						
Printers						
(Other)						

Appendix 12: Resources Required Over Time

Resources Required Over Time (Consolidated)						
Function / Resources	24 hours	48 hours	72 hours	1 week	2 weeks	1 month
All team functions						
Staff						
Area size						
Desks						
Chairs						
Telephones						
Faxes						
PCs						
Printers						
(Other)						
List only the increased amounts in each column. For example the team needs 35 people overall. They assign 15 at the 24 hours slot, another 5 in the 48 hours slot, and 15 more in the 72 hours slot.						

Reprinted with permission from the Disaster Recovery Journal. Available at: http://www.drj.com/new2dr/recoveryteams-e_pearce.pdf.

Appendix 13

Travel Request Form

Appendix 13: Travel Request Form

Make additional copies as needed.

This form should be completed by the team leader and given to the Practice Manager.

Name	Destination	Departure Date / /	Departure Time :
Hotel Reservation Yes ( ) No ( ) Rental Car Yes ( ) No ( ) Cash Advance $_____		Departure Date / /	Departure Time :
Name	Destination	Departure Date / /	Departure Time :
Hotel Reservation Yes ( ) No ( ) Rental Car Yes ( ) No ( ) Cash Advance $_____		Departure Date / /	Departure Time :
Name	Destination	Departure Date / /	Departure Time :
Hotel Reservation Yes ( ) No ( ) Rental Car Yes ( ) No ( ) Cash Advance $_____		Departure Date / /	Departure Time :
Name	Destination	Departure Date / /	Departure Time :
Hotel Reservation Yes ( ) No ( ) Rental Car Yes ( ) No ( ) Cash Advance $_____		Departure Date / /	Departure Time :

Reprinted with permission from the Disaster Recovery Journal. Available at: http://www.drj.com/new2dr/recoveryteams-e_pearce.pdf.

Appendix 14

Recovery Boxes

Appendix 14: Recovery Boxes

Team:

Storage Location:

Contact Name:

Box Identification:

Contents	Comments

Box Identification:

Contents	Comments

1. Storage location refers to the name of the offsite storage facility.
2. Contact name refers to the person who coordinates retrieval of recovery boxes.
3. Box Identification refers to the identifying code on the outside of the box.
4. Contents/comments identify the items stored in the box and special concerns such as update/maintenance or shelf life.

Reprinted with permission from the Disaster Recovery Journal. Available at: http://www.drj.com/new2dr/recoveryteams-e_pearce.pdf.

Appendix 15

Critical Resources to Be Retrieved

Appendix 15: Critical Resources to Be Retrieved

Note: Use this form to document the materials that should be retrieved if you are able to enter your facility following the incident and the items are not badly damaged.

Business Unit: _____

Bldg./Floor:	Location on Floor: (e.g; Northwest Corner)	
Items to Be Retrieved	Comments	Condition*
CRITICAL RECORDS:		
EQUIPMENT:		

Items to Be Retrieved	Comments	Condition*
OTHER:		

*Complete "Condition" at the time of the incident.

Reprinted with permission from the Disaster Recovery Journal. Available at: http://www.drj.com/new2dr/recoveryteams-e_pearce.pdf.

Appendix 16

Personnel Location Control Form

Appendix 16: Personnel Location Control Form

Make additional copies as needed.

COMPLETE DAILY
FORWARD TO THE CRISIS MANAGEMENT TEAM

Date: ____/____/____   Completed by: _____

Operations Team

Name	Recovery Location	Phone Number	Work Schedule From	To

Reprinted with permission from the Disaster Recovery Journal. Available at: http://www.drj.com/new2dr/recoveryteams-e_pearce.pdf.

Appendix 17

Status Report Form

Appendix 17: Status Report Form

Use this form to log significant recovery activities.

Make additional copies as needed.

The team leader is required to submit written recovery status reports daily. Submit completed status reports to the Practice Manager. This status report may be submitted handwritten as long as it is legible.

Date: ___/___/___

Time: ___:___ AM/PM

Name: _____

Department: Operations Team

Comments: _____

_____

_____

_____

_____

_____

_____

Conclusions: _____

_____

_____

Reprinted with permission from the Disaster Recovery Journal. Available at: http://www.drj.com/new2dr/recoveryteams-e_pearce.pdf.

Appendix **18**

Activity Schedule

Appendix 18: Activity Schedule

Plan Reviews

Enter the dates when plan reviews were conducted.

Plan Holders	Due Jan 1	Due Jul 1
Team Leader (Name)		
Alt. Team Leader (Name)		
(Name)		
(Name)		
(Name)		
(Name)		

Training/Exercises

Enter the dates and number of participants for each activity. Each exercise type is expected to be conducted at least once per year.

Activity	Date Conducted	# of Participants	Comments
Orientation			
Team Exercise			
Team Leader Ex			
Functional Exercise			

Team Leaders: Attach participant sign-in sheets, evaluations and comments to this sheet.
Send this page to the Practice Manager no later than December 1.

Task	Required Steps	Expected Results	Task Duration
1.			
2.			
3.			
4.			
5.			
6.			
7.			

Critical Function Recovery Tasks

Function name:_____

Reprinted with permission from the Disaster Recovery Journal. Available at: http://www.drj.com/new2dr/recoveryteams-e_pearce.pdf.

Appendix **19**

Guide to Record Retention

Medical Records*

Patient Charts ....Permanently
Patient Charts—Alternative (adults) ....Ten years after the most recent encounter
Patient Charts—Alternative (minors) ....Age of majority plus statute of limitations
Medical Correspondence (to patients, to referrers
about patients, etc.) ....Permanently with chart
X-rays ....Permanently with chart

Other Medical Record Issues:

Patient Requests Transfer

When transferring medical records, the physician should maintain the original record and should transfer only a copy. You may charge the patient a reasonable fee to reflect the cost of the materials used, the time required to prepare the material, and the direct cost of sending the material to the requesting physician.

(Note: this may be determined by state law, e.g., Georgia has such a law which became effective July 2001.)

The obligation to pay for the record rests with the patient or with the third party who has requested the information. Since this is generally an uninsured service, reasonable attempts may be made on the part of the physician to collect the fee in advance. Nonpayment of the fee, however, is not a reason to withhold the information.

Physician Relocates

Physicians relocating their practice may take the medical records with them or leave the records with a designated custodian with an agreement that they will be permitted ready access to them as required in the future on request.

continues

Physician Ceases Practice

If a physician ceases to practice medicine, he or she may be obligated to either transfer their patients' records to another physician at a local address and phone number or notify each patient that their medical records will be destroyed in (state specific) ___ years, unless they collect the records or request a transfer to another physician within 2 years. You may wish to contact your liability insurer for additional guidance.

Medical Records in a Group Practice That Is Changing

Physicians in a group practice setting usually have an arrangement that clarifies ownership of the records and a transferring policy with respect to patient records. Despite the existence of any such arrangements, it is important to note that any physicians in any setting (e.g., solo practice, group practice, hospital, etc.) are ultimately individually responsible for their own patient records. Physicians must be aware that agreements made with their associates do not supersede their responsibility to patients.

Typically, most physicians in a group practice arrangement will have an agreement with their associates that addresses such items as:

- The method for division of medical records upon termination of the practice agreement. This agreement usually specifies a method for determining custody of the medical records.
- Some reassurance that each physician will have reasonable access to the content of the medical records for preparing medico-legal reports, defending actions, or responding to a complaint investigation.
- Often, if no such agreement exists, physicians dissolving their joint practice try to agree on a system to determine who is "the most responsible physician" for each record. For example, the physician who has created the greatest percentage of the entries in a particular patient record may be expected to continue to maintain it.
- While the above-mentioned approach is customary in most group practices, it is not mindful of the patient's needs. See details in "Ask the Patient" below.

Ask the Patient

Members of a group practice must be cognizant of the fact that it is the *patient's privilege* to choose which doctor they wish to maintain their particular patient records and provide continuing medical care, regardless of the existence of an agreement.

A copy (or original) of that patient's records should be transferred and physicians should agree how the cost of copying and transferring records will be divided within the group. In the case of planned group practice dissolution, the cost cannot be charged to the patient.

Unexpected Dissolution of a Group Practice

Unexpected dissolutions of group practices create special difficulties. Ideally, physicians involved should amicable agree on a strategy for informing patients and dealing with the medical records. In the case of a sudden, unforeseen departure of a partner or associate, records should be kept at their present location until the patient directs where they wish to receive their ongoing health care. Reasonable access to medical records must be given to all former partners and associates.

Statutory Requirements

There are some statutory requirements on the keeping of medical records. For example, certain Medicaid/Medicare reimbursement regulations require that the medical records of recipients be available for review for seven years.

Tax and Financial Records**

Record	Retention
Accounts Payable Ledger	Permanently
Accounts Receivables Ledger—Annual	Six years after the due date of the practice tax return
Accounts Receivable Ledger—Monthly	Two years
Bank Statements with canceled checks	Six years after the due date of the practice tax return
Capital Asset Records	Six years after the due date of the practice tax return for the year in which the asset is disposed
Cash Recipients Journals	Six years after the due date of the practice tax return
Check Register	Six years after the due date of the practice tax return
Daysheets	Six years after the due date of the practice tax return
Deeds, Mortgages, and Bills of Sale	Permanently
Deposit Books and Slips	Six years after the due date of the practice tax return
Depreciation Schedules	Permanently
Encounter Forms	Six years after the due date of the practice tax return
Financial Statements—Annual (year end)	Permanently
Financial Statements—Periodic	Two years
General Ledger	Permanently
Income Tax Returns (correspondence and audits)	Permanently
Income Tax Returns (federal and state)	Permanently
Insurance Policies (expired)	Three years
Insurance Policies, Current Accident Reports, Claims, Policies, etc.	Permanently

continues

Appendix 19: Guide to Record Retention

IRA and Keogh Plan Contributions, Rollovers,
  Transfers, and Distributions ....Permanently
Paid Invoice-Expenses ....Six years after the due date
                                              of the practice tax return
Payroll Ledger ....Six years after the due date
                                              of the practice tax return
Payroll Tax Returns ....Permanently
Petty Cash Vouchers ....Three years
Stock and Bond Certificates (canceled) ....Seven years
Vouchers for Payments to Vendors, Employees, etc.
  (includes allowances and reimbursement of employees,
  officers, etc., for travel and entertainment expenses) ....Seven years

Employer

Employee Personnel Records (after termination) ....Seven years
Employment Applications ....Three years
Time Cards and Daily Attendance Reports ....Seven years

Other

Accident Reports/Claims (settled cases) ....Seven years
Correspondence, General ....Two years
Correspondence, Legal and Important Matters ....Permanently
Correspondence, Routine with Customers or Vendors ....Two years
Minute Books of Directors, Stockholders, Bylaws, and Charter ....Permanently
Trademark Registrations, Patents, and Copyrights ....Permanently

*State guidelines vary. Check with a local medical records training program, your professional liability carrier, or your health care attorney.

**Many of these documents are maintained electronically. We recommend downloading this file to a disk or CD for storage, as indicated.

Source: Reprinted with permission from Gates, Moore, and Company.

Index

Note: Figures are indicated with an italicized page-locator; tables are noted with a *t*.

A

Accident reports, retention guidelines for, 102, 220
Accidents, 3
Accountability procedures, disaster plans and, 58
Accountants, communicating with, 55
Account lockout threshold, enabling, 20
Accounts payable information, storage and accessibility of, 97
Accounts payable ledger, retention guidelines for, 101, 219
Accounts receivable aged trial balance by payor/by patient, storage and accessibility of, 97
Accounts receivable ledgers, retention guidelines for, 101, 219
ACD. *See* Automatic Call Distribution
Activity Schedule
  checklists, 83, 214–215
  critical function recovery tasks, 215
  plan reviews, 214
  training/exercises, 214
Administrative functions, post-disaster resumption of, 58
Advanced encryption standard, 115, 117
Adware, 13, 18–19, 146
AES. *See* Advanced encryption standard
Agility, 147
Air conditioning, 126
Air-sampling smoke detection, 30–31
Allen, Woody, 133
All-risk insurance policies, 134, 135
Alternate practice meeting location, selecting, 88–89
Alternative care, for patients, 81
Alternative sites for practices, 143–151
  calculating cost of down time and, 145–149
  development plan for, 149–150
  key questions related to, 144

Amazon.com, 115
American Power Conversion, website for, 61
American Red Cross, 67
Anthony, Brian, 93
Anthrax, 4
Antispyware software, 15, 17–18
Antivirus software, 13–14
Application service provider option, data backup and, 110
Aqueous film-forming foams, 32
Arson, 27
Assets of practice, protecting and recovering, 93–103
Asymmetric encryption, 23
Automatic backup systems, 111
Automatic Call Distribution, voice/data communications form and, 62, *63*
Automobile insurance, 141

B

Bach, Richard, 1
Backup plan creation, 105–122
  assistance for data backup systems, 113–119
  clinical information, 110
  cost of backing up your data, 112
  key questions related to, 106
  other resources for backing up data, 121–122
  patient demographic and insurance information, 109
  practice financial information, 110
  selecting correct data protection solution, 112
  testing plan and, 119–121
  types of data backup, 112–113
  ways to back up important information, 110–111
Backups
  for computers, 10
  of critical patient and financial data, 93

221

Backups—*continued*
  daily, on all critical files, 95–96
  for data, 10–11, 90
  for financial and billing information, 97
Backup sites, guidelines for travel to, 86–87
Backup software logs, reviewing, 96
Backup systems, HIPAA-compliant, 111, 118
Backup websites, 111
Badges, identification, 53
Banking information, storage and accessibility of, 97
Banks
  communicating with, 55
  non-operational, account management and, 57–58
Bank statements, retention guidelines for, 101
Battery backup systems, 61
Beach Surgical Group (Bay St. Louis, MS), 93
Belarc Advisor, 62
Best practices, industry, for fire protection, 37–39
Billing information, backing up, 97
Bills of sale, retention guidelines for, 101, 219
Biological agents, terrorism and, 3, 4
Biological contamination, 3
Biometric access, 111, 117, 118
Bioterrorism, 4
Blankets, on disaster response checklist form, *68*
Blended model, for data backup, 114
Bond certificates, retention guidelines for, 102, 220
Boot files, backup copies of, 60
Botulinum, 4
Building codes, business interruption insurance and, 139
Building information, storage and accessibility of, 97
Buildings
  damage to, 78
  insurance coverage for, 136
  practice location, disaster planning and, 71–72
Buras, Floyd, 105
Business checks, blank, offsite locations and retention of, 97
Business continuity, 45. *See also* Disaster plan preparation, 94

Businesses
  disaster-related closures of, 134, 144
  significant interruptions experienced by, 5, 5–6
Business functions form, critical, *56*
Business impact analysis, 147–148
Business income protection, 138
Business interruption insurance, 133, 137–141

C

Cable, 62
Call centers, hospitals and, 128
Call orders, 51
Capital asset records, retention guidelines for, 101, 219
Carbon dioxide systems, 34–35
Cascardo, Debra, 53
Cash flow analysis, 148
Cash recipients journals, retention guidelines for, 101, 219
CD copying programs, 119
Cell phones
  chargers for, 65
  disasters and use of, 65, 164
Charge to Go, 65
Check registers, retention guidelines for, 101, 219
Check signers, authorized, 57
Chemical agents, 4
Chemical contamination, 3
Childcare, 52
Children, emergency response cards for, 158, 160–161
Chlorine, 4
Cholera, 4
CiperCide, 27
"Clean-desk policy," 81
Client-site server option, data backup and, 110
Clinical information, backup plan for, 110
Clones of data, 113
Clothing, in disaster kit, 158
"Coding White," 125
Coins, in disaster kit, 164
Cold snaps, 3
Combustion, stages of, 28
Communication, hospital disaster planning and, 131
Communications planning
  alternate practice meeting location and, 89
  disaster plan preparation and, 64

Communication system images, 114
Comprehensive insurance policies, 134
Comprehensiveness, in disaster planning, 168
Compression (ZIP) technology, 117
Computer crashes, 4, 9, 15, 17, 108, 119
Computer disasters, preventing, 10
Computer equipment
  disaster plan preparation and, 59–62
  documentation for, 62
  elevating and securing, in case of floods, 61
Computer Equipment and Software Form, 59, 180
Computers
  dependence on, 9, 10
  emergency shut-down procedures for, 58
  insurance coverage for, 41
  protecting from unwanted eyes and cyberthieves, 12–16
  surge protectors with guarantees on, 66
Computer system, resumption activities for, 90–91
Computer workstation inventory, 120
Conference calls, to employees and key contacts, 64–65
Contingency planning, 45
Contingency plans, for emergency housing, 52
Continuous data protection, 113
Control systems, 36–37
Copyrights, retention guidelines for, 102, 220
Cordless phones, disasters and use of, 65
Corporate Headquarters Phone Numbers List, 82, 192
Corporate sabotage, 27
Correspondence, retention guidelines for, 102, 220
Costs, data backup, 112
CPR training, employees and, 69
Crashes, computer, 4, 9, 15, 17, 108, 119
Creditors, communicating with, 55
Crisis situations, protocols for, 150
Critical Business Functions Checklist, 56, 178
Critical documents storage/accessibility, considerations related to, 98–99
Critical resources, retrieval of, practice resumption and, 87–88

Critical Resources to Be Retrieved List, 83, 206–207
Cyanogen chloride, 4
Cyberthieves, 16–17
  keyloggers, 16
  redirectors, 16–17
  remote access, 17

D

Daily attendance reports, retention guidelines for, 220
Damage-related costs, tracking, 79–80
Damage to facilities, insurance claims and, 78, 79
Data
  clones of, 113
  electronic, protection and recovery of, 95–97
  securing on Internet, 107
  storage of, 10, 149
Data backups, 10–11, 90
  assistance for systems, 113–119
  disaster preparedness and, 72
  online, 111
  outsourcing for, 106–107
  types of, 112–113
Data centers, 28, 60
Data communications, disaster plan preparation and, 62, 64–66
DataEraser, 27
Data loss, 10, 57, 108, 121
  fire and impact on, 27–28
  small businesses and, 106
Data protection solutions, selecting, 112
Data removal, 26–27
Data theft, 15, 107
Daysheets, retention guidelines for, 101, 219
Day-zero threats, 14
Decision makers, disaster planning, 73
Deductibles, business interruption insurance and, 138
Deeds, 157
  retention guidelines for, 101, 219
Demographic information for patients, backup plan for, 109
Department of Labor Statistics, 106
Deposit books and slips, retention guidelines for, 101
Depreciation schedules, retention guidelines for, 102, 219
Desktop computers, security systems for, 14
Digital linear tape, 114

Digital records, 107
Digital technology, 107, 108
Disability insurance, 138, 141
Disaster boxes, 47, 119
  checklist for, 50
  for practices, 48
  protecting against loss of, 49
  security issues with, 49
  storage of, 48
  updating, 48–49
Disaster drills, mock, 69
Disaster insurance
  reimbursement window with, 145
  selecting, 134–136
Disaster kits, building, 158–161
Disaster planning
  effective, 168
  for employees, 153–166
  for hospitals, 125–131
  key questions related to, 2–3
  ongoing process of, 41
  technological disasters and, 11
Disaster plan preparation, 43–75
  building evaluation, 71–72
  computer equipment and software, 59–62
  employee knowledge about emergency plans, 69
  employee preparedness at home, 70–71
  getting started, 73–75
  for given scenarios, 47
  goal of, 45–46
  human resources and, 50–53
  key contacts, 53–55
  key questions related to, 44
  miscellaneous resources, 66
    form for, 67
  practice and training related to, 72–73
  practice operations, 55, 57–58
  risk analysis, 46
  vital records, 69
  voice/data communications, 62, 64–66
Disaster plans, 11
  for hospitals, 128–130
  updating, 169
  writing, 74
Disaster recovery facilities, 60
Disaster recovery "fire drills," 119–121
Disaster recovery kits, hard copies of vital records in, 95
Disaster recovery plan, employee contact list and, 51, 51

Disaster Response Checklist Form, 68, 186
Disasters
  actions to take after, 77–81
  categories of, 3–7
  daily, 5
  preventing, 81–82
Disaster shelters, 165–166
Disaster supply kits, 44, 57
Disaster team, appointing, 73–74
DLT. *See* Digital linear tape
Documentation
  for computer equipment, 62
  disaster planning and, 69
  Family Emergency Plan and, 157
  key information, storage and accessibility of, 97–98
  for practice building, 71
Down time, calculating cost of, 145–149
Drills, 150
Driver licenses, 157
Dry chemical fire suppression systems, 32
Dry-powder fire suppression systems, 32
DSL, 62

E

Earthquakes, 3, 47, 107
Eisenhower, Dwight D., 125
Electrical circuits, overloaded, eliminating, 82
Electrical cutoffs, 71
Electrical fires, preventing, 82
Electrical surges, 10, 15, 24
Electricity, 164
Electric switches, knowing location of, 164
Electrocution, water leaks and, 164
Electronic data, protection and recovery of, 95–97
Electronic medical record programs, key contacts and, 55
Electronic medical records, 9
  digitization of, 107
  management of, in disaster-prone areas, 95
  storage and accessibility of, 97
e-mail, 22, 62
e-mail accounts, free, setting up, 62
e-mail service providers, contact information for, 62
e-mail trees, 52
Emergency equipment, disaster plans and, 58

Emergency evacuations, business interruption insurance riders and, 138
Emergency management, practice recovery steps and, 84
Emergency management agencies, 67
Emergency messaging systems, 64
Emergency Operations Center (Bay St. Louis), 126
Emergency Phone Numbers List, 69, 70, 188
Emergency plans, employees and, 69
Emergency preparedness, 45, 94. *See also* Disaster plan preparation
Emergency responders, communicating with, 55
Employee Contact List, *51*, 173
Employee listing, storage and accessibility of, 97
Employee personnel records, retention guidelines for, 102, 220
Employees
  disaster planning and communicating with, 64
  disaster planning for
    disaster kits, 158–161
    disaster shelters, 165–166
    evacuation, 165
    Family Emergency Plan, 155–156
    fire, 165
    food, 161–162
    key questions related to, 154
    pets, 161
    utilities, 163–165
    water, 162
  emergency plans and, 69
  home-based preparedness and, 70–71
  new, preparedness manual and, 72
  practice resumption and, 78
  transitioning daily routines for, after a disaster, 147
Employer records, retention guidelines for, 102, 220
Employment applications, retention guidelines for, 102, 220
Encounter forms, retention guidelines for, 102, 219
Encryption, 10, 22–24, 111
  asymmetric, 23
  safe backup systems and, 115
  symmetric, 23–24
Encryption keys, 111
  creating, 115
  with eSureIT, 116, 117

EOC. *See* Emergency Operations Center (Bay St. Louis)
Equipment damage, 149
  temporary office space and, 80
Equipment specifications inventory/listing, 99
Equipment storage companies, offsite, 99
Errors
  data loss and, 108
  disaster plan preparation and, 46
  fire hazards and, 40
Escape routes, 71
Essential Cosmetic Surgery Companion, The: Don't Consult a Cosmetic Surgeon Without This Book (Kotler), 1
eSureIT, 116–118
Evacuation planning, 58, 69
Evacuation routes, for families, 158
Evacuations, 165
  to emergency shelters, 166
  from hospitals, 126
Evans, Harris, 77
External hard drive products, 90
Extra expense insurance, 141

F

Facial recognition, 111, 118
Facility size, alternate practice meeting location and, 89
Family Emergency Plan, *155–156*
Federal Trade Commission, 15
File cabinets, miscellaneous resource form and, 66, *67*
File transport protocol, firewalls and, 21–22
Financial accounts, cyberthieves and, 16, 17
Financial information, backing up, 97, 110
Financial needs, of employees, 52
Financial records and statements, 157
  retention guidelines for, 101–102, 219
Fingerprint recognition, 111, 118
Fire department, 55, 67
Fire detection system types, 29–31
  air-sampling smoke detection, 30–31
  intelligent spot-type smoke detection, 30
  linear-thermal detection, 31
  spot-type smoke detection, 29–30
Fire extinguishers, 33, 71, 72, 157
  disaster response checklist form, *68*

Fire extinguishing methods, 28
Fire insurance, 139, 140
Fire protection solutions, choosing, 28–29
Fires, 3, 107, 144
　computer failures and, 90
　data loss and, 27–28
　leaving home in case of, 165
Fire suppression systems, 71
　types of, 31–35
　　dry chemical, 32
　　fire extinguishers, 33
　　fluorine-based compounds, 35
　　foam, 32
　　gaseous agents, 34–35
　　total flooding fire extinguishing systems, 33–34
　　water mist systems, 32–33
　　water sprinkler systems, 32
Fire Triangle, 28
Firewall drives, 61
Firewalls, 21–22
　hardwired, 96
　soft-wired, 96
First-aid kits, 44, 158
First-aid supplies, on disaster response checklist form, 68
First-aid training, for employees, 69
Flame detectors, 29
Flaming fire stage of combustion, 28
Flat dollar amount, deductibles, business interruption insurance and, 138
Flexibility, in disaster planning, 168
Flood insurance, 135
Floods, 3, 47, 107, 144, 149
　computer failures and, 90
　named peril insurance policies and, 135
　safeguarding computer equipment and, 61
Floors, raised, 37, 40
Fluorine-based compounds, 34, 35
Foam, fire suppression with, 32
Folino, Frank, 128, 130
Food
　in disaster kits, 158, 161–162
　on disaster response checklist form, 68
Franklin, Benjamin, 93
Freezing weather, 47
FTP. *See* File transport protocol
Furniture, on miscellaneous resource form, 66, 67

G

Gas, natural, 163
Gaseous agents, fire extinguishing with use of, 34–35
Gas main valves, 71
GE Medical Buildings, 99
General ledgers, retention guidelines for, 102, 219
Generators, 147
　hospitals and, 126
　multi-KV, 69
Geographic factors, disaster plan preparation and, 46
Google Groups, 19
Google Toolbar, 19
Group practice
　changing
　　medical records in, 100–101
　　retention of medical records and, 218
　unexpected dissolution of, records retention for, 101, 218–219
Guide to Record Retention Checklist, 83
Gutman, Sam, 105

H

Hackers, 22
Halon 1301, 34
HAM radio setups, 126
Hard drives
　erasing, 26–27
　failures of, 108
Hardware
　failure of, 90
　theft of, 108
Hardware information, documenting, 95
Hardware licenses, updating, 62
Hardwired firewalls, 96
Hazardous materials, 71
Health department, communicating with, 55
Health information, spyware and breaches of, 15
Health Insurance Portability and Accountability Act, 45, 107, 108
Hearing impaired persons, 160
Heat detectors, 29
Heat waves, 3
HFC-236fa fire extinguishers (FE 36), 33
HIPAA. *See* Health Insurance Portability and Accountability Act

Hirsch, Leslie, 127, 128
Historical factors, disaster plan preparation and, 46
Home-based preparedness, employees and, 70–71
Homeowners' insurance, 141
Home safety, employees and, 154, 157
Hospital perspective, 125–131
 before a disaster, 128–129
 key questions related to, 127
 what hospitals can do for you and your practice, 130–131
Hospitals, communicating with, 55
Hotmail, 62
Housekeeping, disaster prevention and, 81
Housing, emergency, contingency plans for, 52
Human error, disaster plan preparation and, 46
Human resources, disaster plan preparation and, 50–53
Hurricane Katrina, ix, x, xi–xii, 43, 58, 77, 93, 105–106, 143, 150, 154
 business shutdowns and, 137
 hospital operations and, 125–131
 insurance and, 136
 law offices and, 140
Hurricanes, 3, 144
 computer failures and, 90
 named peril insurance policies and, 135
Hurricane Wilma, 95
Hydrogen cyanide, 4

I

Identification badges, 53
Identity theft, spyware and, 16
Incipient stage of combustion, 28
Income protection, business interruption insurance and, 139, 140
Income tax returns, retention guidelines for, 102, 219
Independent Insurance Agents and Brokers of America, 6
Industrial accidents, 3
Inergen, 34
Inert gases, 34, 35
Information security procedures, disaster prevention and, 82
Information system capabilities, resumption activities for, 90–91

Information technology experts, data backup system assistance and, 113–114
Injured employees, contacting, 52
Inspections, 72
Institute for Business & Home Safety, 75, 134
Insurance, 169
 business interruption, 137–141
 for computers and other technologies, 41
 disaster, selecting, 134–136
 extra expense, 141
 flood, 135
 gaps in coverage, 6
 key questions related to, 134
 property, 136–137
 reviewing current coverage, 136
Insurance claims, filing, 78, 79
Insurance companies, communicating with, 55
Insurance information, 157
 patient, backup plan for, 109
Insurance policies, 94
 retention guidelines for, 102, 219
 storage and accessibility of, 98
Intelligent spot-type detection, 30
Intense heat stage of combustion, 28
Internal Medicine of Long Beach (Mississippi), in wake of Hurricane Katrina, 77
International Communications Research, 6
Internet, 40, 147, 149
 antispyware legislation and, 18
 backup for, 118
 dial-up access to, 62
 firewalls and, 21
 securing data on, 107
 viruses and, 12
Internet-based management information systems, 95
Internet log-on codes and passwords, up-to-date copies of, 60
Internet service providers
 password availability and, 97
 storage and accessibility of information on, 98
Intranets, firewalls and, 21
Intronis Technologies, 116
Inventory, 149
Ionization detectors, 30

IRA information, retention guidelines for, 102, 220

J
Job continuity, 53

K
Kellum, Ron, 43
Keogh plan contributions records, retention guidelines for, 102, 220
Key contact information, storage and accessibility of, 98
Key contacts, disaster plan preparation and, 53–55
Key Contacts List, 54, 175
Keyloggers, 16
  Web sites and hosting of, 17
Kotler, Robert, 1

L
Laboratory results, 9
Landscaping, video taping, property documentation and, 137
Laptop computers
  backing up data on, 96
  disaster plans and, 60
  security systems for, 14
  theft of, 108
Lawrence, Matt, 153
Lawyers
  communicating with, 55
  Hurricane Katrina and, 140
Lay offs, 148
Leftwich, Hal, 125, 126, 127
Legislation, antispyware, 18
Lewisite, 4
Liability insurance, 136, 138, 169
Life insurance, 141
Lightning, 27
Linear tape-open, 114
Linear thermal detection, 30, 31
Linux, antivirus software and, 14
Local area networks, viruses and, 12
Locks, for practice building, 72
Locum tenens employment, 144
Loss, data, 10, 57, 108, 121
LTO. *See* Linear tape-open

M
Macintosh computers
  antivirus software and, 14
  System Profiler and, 62
Mackay, Harvey, 167

Mac OS X, 14
Mail service, rerouting, 78–79
Malpractice insurance, proof of, 157
Man-made disasters, 4
Marriage licenses, 157
McAfee, 14
Meals ready to eat, 161
Media, communicating with, 55
Medical correspondence, retention guidelines for, 100, 217
Medical equipment, emergency shut-down procedures for, 58
Medical licenses, 157
Medical records
  in changing group practice, retention guidelines for, 100–101, 217
  loss of, after Hurricane Katrina, 105
  statutory requirements on retention of, 101
Medical records management, practice disruption and, 94
Medical school diplomas, 157
Medical staff recovery, hospitals and, 127
Medical trailers, mobile, 143
Medications
  on disaster response checklist form, 68
  key contacts list and, 54, 54
Microchipping, of pets, 161
Microsoft Office, 90
Microsoft Windows
  antispyware software and, 15
  antivirus software and, 14
Minute books of directors, stockholders, bylaws, and charters, retention guidelines for, 102, 220
Miscellaneous resources form, 184
Miscellaneous resources form, 66, 67
Mission-critical facilities, protecting, 37
Mobile covered trailers, 99
Mobile medical trailers, 143
Mobile recovery box, 87
Mobile recovery services, features of, 147, 151
Modem, voice/data communications form and, 62, 63
Monitoring software, 15–16
Moran, Kate, 130
Mortgages, retention guidelines for, 101, 219
MREs. *See* Meals ready to eat
MSN Toolbar, 19
Multi-KV generators, 69
Mustard gas, 4

Index | **229**

N

Named peril insurance policies, 134–135
National Fire Protection Agency, 27, 35
National Flood Insurance Program, 135
National Guard, 130, 154
National Institute of Standards and Technology, 117
National Security Agency, 117
Natural disasters, 3, 47, 107, 150
 disaster plans for employees, 154
 named peril insurance policies and, 134–135
 risk factors for, 48
Natural gas safety and shut off valves, 163, *163*
Nerve agents, 4
NetWare, antivirus software and, 14
Network Appliance storage systems, antivirus software and, 14
Networks, implementing security on, 96
New Orleans *Times-Picayune*, 130
News media statements, 84
NFIP. *See* National Flood Insurance Program
Norton Utilities, 14
Nuclear accidents, 3

O

Office closures, temporary, procedures for, 78–79
Office Depot, 49
Office furniture, miscellaneous resource form and, 66, *67*
Office space, temporary, 99
Offsite equipment storage companies, 99
Offsite storage locations, identifying, 96
Offsite stored materials, practice resumption plan and, 87
One-stop insurance purchasing, 136
Online backup
 advantages with, 114–115
 of data, 111
 systems for, 114
Open for Business® toolkit, 75
Open peril insurance policy, 134
Operating systems, backup copies of, 60
Operational needs, alternate site development plan and, 149
OrangeBoxx, 50
Orientation sessions, for new employees, disaster preparedness and, 73
Out-of-business coverage, 138

Out-of-town contact person, Family Emergency Plan and, 154, *155*, 157, 158, 161
Outsourcing, for data backup, 106–107

P

Packet filtering, 22
Paid invoice expenses, retention guidelines for, 102, 220
Paper supplies, on disaster response checklist form, *68*
Passports, 157
Password history, setting up, 20
Passwords, 10
 commonly used, examples of, 19–20
 computers protected with, 19–21
 cyberthieves and, 16, 17
 documenting, 62
 good examples of, 20–21
 secure lists of, 97
Patents, retention guidelines for, 102, 220
Pathology reports, digital, 107
Patient charts, retention guidelines for, 99, 217
Patient demographic and insurance information, backup plan for, 109
Patient lists
 practice resumption plan and, 83
 storage and accessibility of, 98
Patient privilege, 218
Patient records, key contacts list and, 54, *54*
Patient requests transfer records, retention guidelines for, 100, 217
Patients
 alternative care for, 81
 disaster planning and communicating with, 51, 64
 disasters and impacts on, 46
 key contacts, disaster plan preparation and, 53–55
 practice resumption plan and, 80–81
 recovery time objective and, 148
Payroll continuity, 53
Payroll ledgers, retention guidelines for, 102, 220
Payroll periods, business impact analysis and, 148
Payroll tax returns, retention guidelines for, 102, 220
PBX. *See* Private Branch Exchange
Peril insurance policies, 134
Peripherals, documentation for, 62

Personal digital assistants, 52, 96
Personnel, contacting, practice resumption plan and, 85
Personnel Location Control Form, 83, 210
Persons with disabilities
  emergency backup for, 160
  emergency response cards for, 158
Pets
  disaster plans and, 161, 166
  hospitals, disasters and, 131
  water for, 162
Petty cash vouchers, guidelines for retention of, 102
PHA. *See* Protected health information
Pharmaceuticals, offsite storage of, 99
Phone chargers, portable, *66*
Phone trees, 44, 52
Phosgene, 4
Photoelectric detectors, 29
Physical facilities, disaster plan preparation and, 46
Physical security procedures, disaster prevention and, 82
Physician ceases practice information, retention guidelines for, 100, 218
Physician relocation information, retention guidelines for, 100, 217
Physicians
  alternative sites for practices, 143–151
  business interruption insurance for, 137–141
  disaster planning for, 43
  on hospital disaster teams, 129
Picture archiving, 114
Plague, 4
Plain old telephone service, 65
Player, Gary, 43
Point-in-time copies, 113
Police department, communicating with, 55
Pop-up blockers, 19
POTS. *See* Plain old telephone service
Power backups, 61
Power outages, 69, 146, 149, 150, 164
Power strip surge protectors, levels of power for, 25
Power surges, 4, 24, 27
Practice assets
  protecting and recovering, 93–103
  key questions related to, 93–94

Practice operations, disaster plan preparation and, 55, 57–58
Practice recovery site, guidelines for travel to, 86–87
Practice recovery steps, 84–86
Practice Recovery Work Area Checklist, 82, 194
Practice resumption plan, 77–91
  actions to take after a disaster, 77–81
  checklists, 82–83
  critical resources to be retrieved, 87–88
  guidelines for travel to practice recovery site, 86–87
  implementing, 81
  importance of, 91
  key questions related to, 77–78
  offsite stored materials, 87
  patient list, 83
  practice recovery steps, 84–86
  prevention, 81–82
  resumption activities for information system capabilities, 90–91
  training/exercise schedule, 88–90
Practices, finding alternative site for, 143–151
Preparedness, practice resumption plan and, 88
Preparedness manual, new employees and, 72
Prescription information, 157
Pretty Good Privacy, 23
Prevention, disasters and, 81–82
Printers, documentation for, 62
Private Branch Exchange, voice/data communications form and, 62, 63
Private keys, 23
Profit protection insurance, 138
Programmable call forwarding, 64
Pro-Inert, 34
"Proof of loss," in insurance disputes, 137
Property insurance, 136–137, 138
Property protection plan, 47
Protected health information, backing up, 108–109
Protocols, for crisis situations, 150
Proxy service, 22
Public keys, 23
Pull stations, 35

Q

Q fever, 4
QuickBooks, 90, 97

R

Raised floors, 37, 40
RBackup Remote Backup Software, 121–122
Record retention, guide to, 93, 99–102, 217–220
Records storage companies, 99
Recovery box, 87
Recovery Boxes List, 83, 202–203
Recovery services, mobile, 147, 151
Recovery teams, for hospitals, 129–130
Recovery time objective, 148
Redirectors, 16–17
Redundant secure data centers, 117
Referring physicians, practice resumption plan and, 83
Referring physicians listings, storage and accessibility of, 98
Remote access spyware, 17
Remote storage
 computer system backups and, 90
 of reports, 91
Resource form, miscellaneous, 66, 67
Resources Required Over Time List, 83, 196–198
Retrieval, of backup files, 118
Revenue projections, 148
Riot-control gases, 4
Ripkin, Cal, Jr., 77
Risk assessment, alternate site development plan and, 149
Risk Assessment Questionnaire for your Practice, 49t
Risk factors, disaster plan preparation and analysis of, 46
Risk management, 94

S

Safety coordinators, 69
Sarin, 4
Satellite telephones, 126, 130, 146–147
Scans, virus, 13
Secure Sockets Layer, 111, 115, 117
Security, 149
 clearances, 53
 codes, 60
 disaster boxes and, 49
 practice building, 71–72
 software, 13–16
Security suite, ideal, 14
Seniors
 emergency backup for, 160
 emergency response cards for, 158
September 11, 2001 terrorist attacks, xii, 4, 6
Servers
 implementing security on, 96
 security systems for, 14
Sewer lines, 71
Shakespeare, William, 143
Sharp, Robert, 143
Shelving, temporary, miscellaneous resource form and, 66, 67
Sick employees contacting, 52
Signaling devices, 36
Simplicity, in disaster planning, 168
SkillPath, 4
Small businesses
 disaster-related closures of, 134, 144
 inadequate insurance coverage for, 6
Smoke detection, variables related to, 31
Smoke detectors, 29, 157, 165
Snapshot technology, 113
Snow, C. P., 9
Social security cards, 157
Software
 adware, 13
 antispyware, 15, 17–18
 backup copies of, 60
 disaster plan preparation and, 59–62
 documentation for, 62
 monitoring, 15–16
 security, 13–16
 spyware, 13
Software information, documenting, 95
Software licenses, updating, 62
Soft-wired firewalls, 96
Sophos, 13, 14
Source codes, 60
South Alabama Medical Center, 126
Spam, 146
Speakerphones, 64
Spikes, 24, 26
Spot-type smoke detection, 29–30
Sprinkler system failures, 4, 149
Spyware, 13, 15, 108
 behavioral change and, 18
 legislative change and, 18

Spyware—*continued*
  protecting against, 18–19
  technological responses to, 17–18
SSL. *See* Secure Sockets Layer
Staff, emergency procedures and, 168. *See also* Employees
Staphylococcal enterotoxin B, 4
Stark Laws, 128
Stateful inspection, 22
Status Report Form, 83, 212
Statutory requirements, medical records retention and, 101, 219
Stock certificates, retention guidelines for, 102, 220
Storage
  of data, 149
  of disaster boxes, 48
  of key information, 97–98
  of key practice assets, 94
  Recovery Boxes List, 202–203
Storm drains, 71
Subscriptions, antivirus software, 15
Supplies, offsite storage of, 99
Support, for antivirus products, 14
Surge protection, levels of, 24–26
Surge protectors, 96
  with guarantees on computers, 66
  response time with, 66
  shopping for, 26
  Underwriters Laboratories ratings for, 65
Surges, 24, 26
Surge stations, 25
Symantec, 14
Symmetric encryption, 23–24
System Profiler, 62

T

Tape backup, 113, 114
Tate, William, 126
Tax records, retention guidelines for, 101–102, 219–220
Team leaders, practice recovery steps and, 84
Team members, practice resumption plan and information given to, 85–86
TechAtlas, 119
Technological disasters, 9–41
  adware and spyware, 13
  bottom line, 40–41
  common mistakes related to, 40
  control systems, 36–37

raised floors, 37
cyberthieves, 16–17
  keyloggers, 16
  redirectors, 16–17
  remote access, 17
data removal and erasing hard drives, 26–27
encryption, 22–24
  asymmetric, 23
  symmetric, 23–24
fire, impact on data loss and, 27–28
fire detection system types, 29–31
  air-sampling smoke detection, 30–31
  intelligent spot-type smoke detection, 30
  linear thermal detection, 31
  spot-type smoke detection, 29–30
fire protection solution choices, 28–29
fire-suppression system types, 31–35
  dry chemical, 32
  fire extinguishers, 33
  foam, 32
  gaseous agents, 34–35
  total flooding fire extinguishing systems, 33–34
  water mist systems, 32–33
  water sprinkler systems, 32
firewalls, 21–22
fluorine-based compounds, 35
industry best practices, 37–39
key questions related to prevention of, 10
passwords to protect computers, 19–21
protecting mission-critical facilities, 37
pull stations, 35
responses to spyware, 17–18
  behavioral change, 18
  legislative change, 18
  technological, 17–18
security software, 13–16
signaling devices, 36
spyware and adware, protecting against, 18–19
surge protection, levels of, 24–26
varieties of, 11
viruses, 12
Telephones
  Corporate Headquarters Phone Numbers List, 192
  disaster plans and, 164–165
  Emergency Phone Numbers List, 188

Temporary office space, 94, 99
Terrorist attacks, 144, 150
Terrorists, 3-4
Testing, backup plans, 119-121
Theft, 1, 149
  computer, 108
  data, 15, 107
Time cards, retention guidelines for, 102, 220
TK8 backup, 122
T1 lines, contingency plans and, 61
Tool kit, on disaster response checklist form, 68
Tornados, 3, 135, 144
Total flooding fire extinguishing systems, 33-34
Touro Infirmary (New Orleans), 127, 128, 129, 130
Trademark registrations, retention guidelines for, 102, 220
Trailers, mobile covered, 99
Training/exercise schedule, practice resumption plan and, 88-90
Transient voltage, 24
Transient voltage surge suppressor, 65
Transportation, of employees, in wake of disaster, 52
Transportation accidents, 3
Travel bags, 166
  items in, 159-160
Travel Request Form, 83, 200
Trojan horses, 12
Trusted Choice, 6
Tulane University Hospital and Clinic, 131
Tularemia, 4

U

Underwriters Laboratories (UL) ratings, for surge protectors, 26, 65
Uninterruptible power supplies, 25, 61, 96
Unix, antivirus software and, 14
UPSs. *See* Uninterruptible power supplies
User names, cyberthieves and, 16, 17
Utilities, 157, 163-165
  communicating with company staff, 55
  electricity, 164
  gas, 163
  phone, 164-165
  shutoffs for, 71
  water, 164

Utility disruptions, business interruption insurance riders and, 138

V

Vendors
  alternate site development plan and, 150
  business impact analysis and, 149
  contact information for, 60, 98
  disaster planning and communicating with, 64
  practice resumption plan and, 83
  vouchers to, retention guidelines and, 220
  workplace recovery centers and, 145
Venezuelan equine encephalitis, 4
Very early smoke detection (VESD), 30, 31
Veterans Association, 107
Video taping, property documentation and, 137
Violence, intentional acts of, 3
Viral hemorrhagic fevers, 4
Virtual private networks, 107
Viruses, 12, 15, 108
Virus protection, 10
Virus scanners, 13
Visible smoke stage of combustion, 28
Vital instruments, offsite storage of, 99
Vital records
  disaster planning and, 69
  identifying and accessing, 95
  protection of, against all disasters, 94
Vital Records List, 71, 189
Vital records program, current, assessment of, 94
Vital records program custodians, designating, 94
Voice/data communications, disaster plan preparation and, 62, 64-66
Voice/Data Communications Form, 63, 182
Voicemail, 64
  voice/data communications form and, 62, 63
Voice messages, practice resumption and, 79
Volcanic eruptions, 3
Vouchers, retention guidelines for, 102
Vulnerabilities
  alternate practice meeting location and, 89
  reducing, 167

W

Waiting time, deductibles, business interruption insurance and, 138
Walkie-talkies, 127
WalMart, 49
Water
  in disaster kits, 158, 162
  on disaster response checklist form, *68*
  hospitals, disaster planning and, 126
  purifying, 162
Water leaks, 164
Water main valves, 71
Water mist systems, 32–33
Water pipe breaks, 4
Water sprinkler systems, 32
Website hosting company, contact information for, 62
Websites
  backup, 111
  hospital, 129
  keylogger hosting and, *17*
Wheelchair-bound persons, 160
Whistles, on disaster response checklist form, 68
Wildfires, 47
Wills, 157
Wind, 47
Windows computers, Belarc Advisor and, 62
Workplace recovery centers, 145
World Trade Center, September 11, 2001 terrorist attacks on, 4, 6
Worms, 12, 15
W-2 forms, 157

X

Xenon flash tubes, 36
X-ray reports, retention guidelines for, 217
X-rays, digital, 107
X-ray studies, 9

Y

Yahoo, 62